D1029944
Ex Libris
Deirdre Kumpfbeck

THE DESIGN AND CONSTRUCTION OF STABLES AND ANCILLARY BUILDINGS

The Design & Construction of

STABLES

and Ancillary Buildings

PETER C. SMITH

Associate of the Royal Institute of British Architects

WITH SEVENTEEN DETAILED LAYOUTS
FIFTEEN DIAGRAMS AND
ONE HALFTONE ILLUSTRATION

J. A. ALLEN & CO LTD

LONDON

J. A. ALLEN & CO LTD
1 LOWER GROSVENOR PLACE
LONDON S.W.1

FIRST PUBLISHED 1967

REPRINTED 1973, 1975, 1979

SBN 851310001

Printed in Great Britain by
Lewis Reprints Ltd.
member of Brown Knight & Truscott Group
London and Tonbridge

Contents

Page

Illustrations

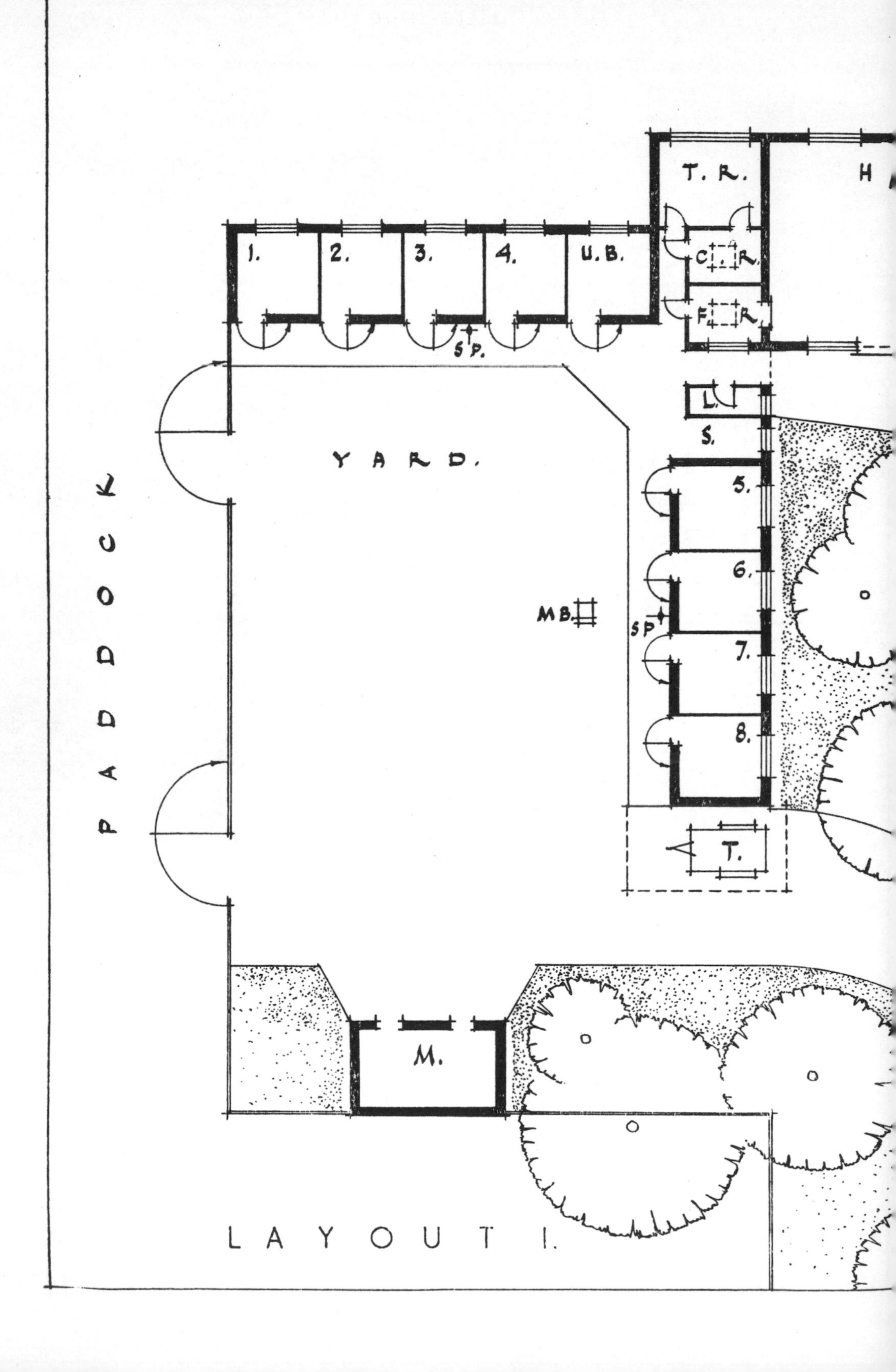

LAYOUT I.

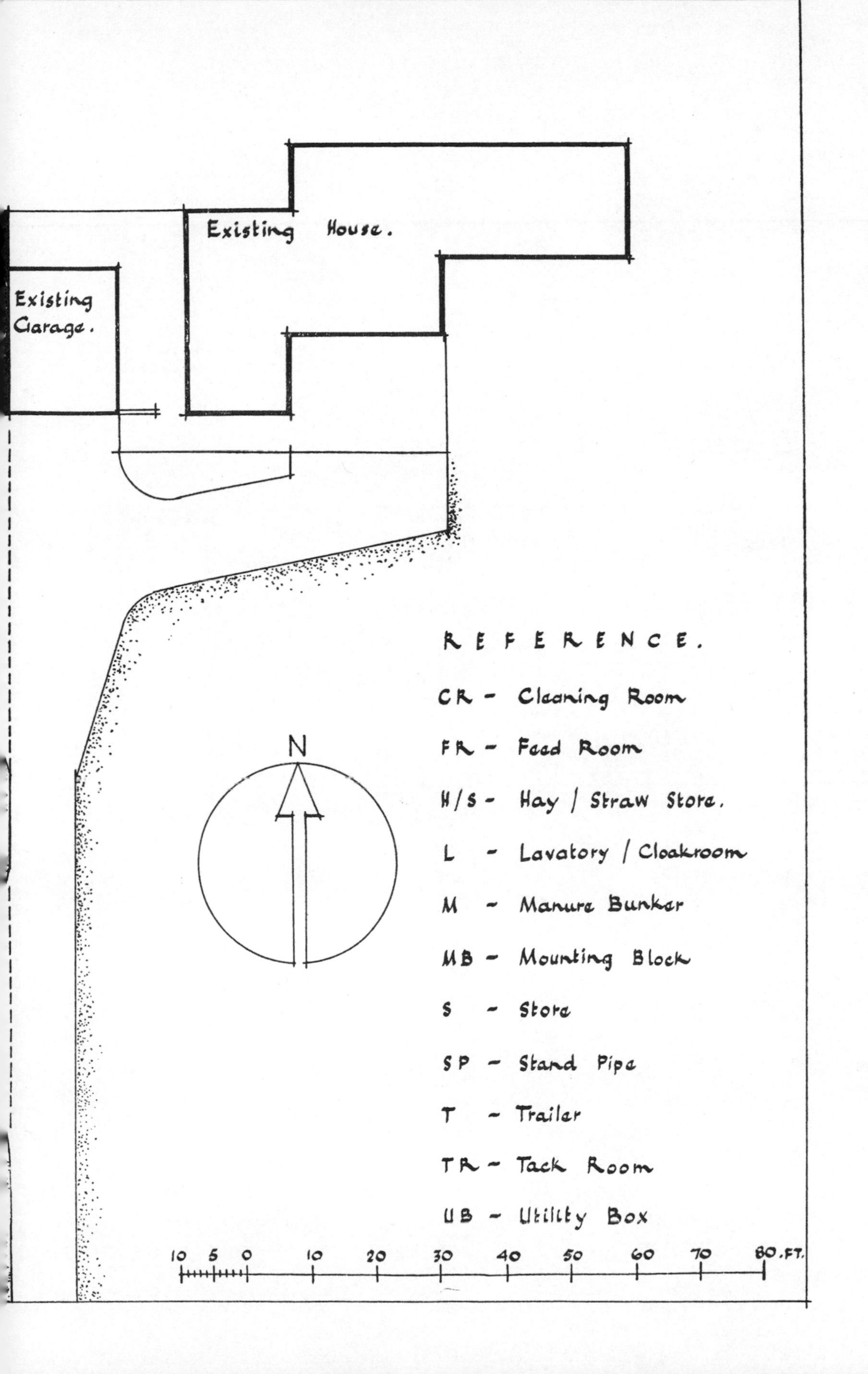

Existing House.
Existing Garage.
REFERENCE.
CR - Cleaning Room
FR - Feed Room
H/S - Hay / Straw Store.
L - Lavatory / Cloakroom
M - Manure Bunker
MB - Mounting Block
S - Store
SP - Stand Pipe
T - Trailer
TR - Tack Room
UB - Utility Box
N
10 5 0 10 20 30 40 50 60 70 80.FT.

INTRODUCTION

It is perhaps the good fortune of few persons to be asked to write a book on a subject which combines the interests of both the business and relaxation sides of their lives. In spite of this, I must admit to a certain trepidation of approach in my initial writings. Any subject relating to horses is surrounded by many opposing theories and to produce a book capable of satisfying all readers would be a practical impossibility. This is, I am advised, the first book written which is entirely devoted to the subject of stabling and I have therefore no opportunity to learn from the mistakes of previous authors on this subject.

In these circumstances, I decided to base the book on my own experience both as a practising architect and, I hope, a practical horseman. It soon became evident however that my own knowledge of stable management was insufficient to deal adequately with the extensive field of subjects which needed to be covered in this book, if the final result was to be of the standard I desired. I had therefore recourse to consult with friends and other authorities so that my thoughts might be safely conducted along the right lines. The horse, fortunately, forges a close link between those who spend much time in his company and I have received the most enthusiastic and sympathetic assistance from everyone I have consulted. My most grateful thanks to them is recorded at the end of this introduction.

The interest in the horse as a means of exercise, pleasure and indeed of medical treatment has progressively increased since the end of the second world war. Stables however are for the most part old buildings, many of which were erected during the latter half of the last century. The maintenance costs of such buildings are high and at the same time one must ask the question, are they suitable for the purpose? Most stable buildings appear to me to have happened, rather than been designed. The layout of many of the older buildings which were erected in the days when grooms were both cheap and plentiful give cause for serious re-assessment when they have to be run, through economic considerations, with half the number of the original staff or even less. At the same time the grooms in those days were men, whereas nowadays most stable staff is composed of young girls who are less able to carry heavy loads considerable distances.

Many owners will probably require professional advice and will employ the services of an architect. The housing needs of any animal requires expert knowledge, both theoretical and practical and such knowledge cannot be gained in a short time and without practical experience of the animal concerned. When an architect is consulted, he is expected to produce sketch plans within a few weeks, far too short a time to gain the knowledge required. Some owners may wish to design their own buildings but may lack the necessary technical knowledge to enable them to take their ideas beyond the layout stage. This will entail them placing their designs in the hands of a builder, who, without previous experience, will not necessarily fully appreciate the constructional and service problems relating to this type of building. I have therefore written this book both as a reference for the architect and also for the use of anybody wishing to erect stables, or to alter or adapt existing buildings.

In considering the architect's needs, I have arranged the contents to suit his design and constructional requirements in accordance with normal practice. Part 1 covers his requirements at sketch plan stage and Part 2 his requirements at working drawing stage.

There are many specialist establishments in the country, covering breeding, racing, etc. The requirements of such establishments differ considerably from those of hunting and hacking stables which has been my concern in this book, and would require a separate work to each of them.

This book has been written for essentially two professions whose lives seldom converge. I have avoided as far as possible the use of technical language. It is thus hoped that the contents will be the easier comprehended by both parties and that the technicalities relating to both horses and structure will have been adequately explained.

Before completing this introduction, please let me make it clear that although certain preferences are expressed throughout the book, these are personal. It is fully appreciated that there are many opinions on all matters relating to stabling, to the care of the horse, and to the composition and furnishing of the stable group. I have therefore, tried to incorporate all the items that any design may require. There will seldom be a stable built which will include the entire contents of this book or all the refinements described. I feel however, that it would be unwise to omit any information which may be of use thereby leaving an incomplete and unsatisfactory work of reference.

Finally I wish to express my grateful thanks to Mr. Wm. C. Miller, F.R.C.V.S., F.R.S.E., who provided much information and assistance; to Mr. and Mrs. V. C. Ellison for their very considerable help and encouragement, and to Mr. A. R. L. Escombe, M.F.H., for much useful advice.

In the related specialist spheres, I extend my thanks to Mr. Ian H. Duff, T.D., M.I.H.V.E., and to Mr. J. L. Wallace, B.SC.ENG., A.C.G.I., M.I.H.V.E., of Messrs. Rosser & Russell Ltd., Heating and Ventilating Engineers for their advice on the chapter dealing with ventilation and in particular for the scheme showing a mechanical ventilation system, to Mr. S. A. Rees, the Secretary of the Federation of Building Block Manufacturers for information on blockwork and to Mr. W. H. Richardson, M.A.S.S.E., of E. Wright & Co. Ltd., Electrical Engineers, for his advice and assistance on the chapter dealing with the electrical installation.

I also wish to express my thanks for their help to the following authorities and firms, The Chief Officer of the Hertfordshire County Fire Brigade for his advice relating to the chapter on fire precautions, to Messrs. Celotex Ltd., for information dealing with fire resisting ceilings and for their Vapour-Check Insulation Board, to the Cement Marketing Company Ltd., for information on cement products, to Messrs. Charles Clark (Commercial Vehicles) Ltd., for information on motor horse boxes; to Messrs. R. A. Lister Ltd., for information on bulk storage hoppers; to Messrs. Cooper, MacDougal & Robertson Ltd., for information and advice on the control of flies; to Messrs. Mabbott & Co. Ltd., for details of stable and saddle room fittings, and to Messrs. Rice Trailers Ltd., for information on horse trailers.

PULRIDGE HOUSE,
LITTLE GADDESDEN,
HERTFORDSHIRE.

January, 1967.

1 THE FUNDAMENTAL REQUIREMENTS OF PLANNING THE STABLES AND THE ANCILLARY BUILDINGS

Reasons for the provision of stables

Before considering the planning and construction of stables in detail, a greater appreciation of the fundamental requirements might be gained, by considering briefly the reasons for providing stables. A horse living out in its natural surroundings has no need of protection from the weather providing it has a sufficiency of good food. The thickness of the coat during the winter months, the natural oils in the skin and the fact that it may move about freely, are all adequate protection from inclement conditions. A horse living under such conditions, however, is fit only for the lightest work.

A horse in full work needs to be fit and to be kept clean. To obtain this condition requires the removal of surplus fat, a clipped coat and regular grooming. Thus not only is the coat lightened but cleaning and grooming take away a large proportion of its natural protective oils. A horse in this condition cannot satisfactorily live out and artificial methods of protection must be provided, by means of blankets and stables. Stables have, therefore, to provide to a great extent the protection to the animal which has been removed by the requirements of work and cleanliness.

Individual requirements of stable buildings

It must be appreciated that the requirements of no two owners will be the same. Discounting the differences in sizes and accommodation needs of the various establishments and the purpose to which they may be put, it will be found that many owners have fixed ideas on the running of their stables which may materially affect the design. It is important therefore to discuss with the owner, not only his accommodation needs but his routine to be carried out daily. Information on the furnishing needs of each unit will also be required.

Some days spent observing the routine in a stable will help to give the uninitiated a clearer idea of the requirements, better still spend some part of one's holiday working in a stable of similar type and size to the one to be designed. The insight into the routine and needs will be the more appreciated. Never hesitate to ask questions of the client, however simple they may seem. Horsemen are always happy to talk on their subject so no excuse can be given for not obtaining adequate information for design purposes.

Many stables will be erected as an adjunct to an existing house and the accommodation of the buildings will depend on the ancillary accommodation for both staff and storage already available on the site. If the buildings are to be erected in conjunction with a new house, full staff and storage accommodation may be needed within the stable buildings. The provision of storage will also depend and vary on the method of running the stables; if for instance they are to be attached to a farm, the farmer might grow his own hay and leave it in the stacks to be cut as required. In such a case the provision of a hay store would not be necessary or would be of minimum size, say enough for one or at the most two weeks supply. Such possibilities are dealt with under their respective headings later in this section.

Staff accommodation will vary enormously. Most large establishments will have some staff living in and the provision of bed-sitting rooms or flatlets may be needed. Many small establishments are run by the client and his family, with help from local girls who will travel to and from the stables each day. Small establishments will seldom require any staff accommodation as the grooms will either be working in the stables or exercising the horses during the working day. When staff accommodation is required it should be placed preferably near to the stable buildings.

Principal requirements of stables

The basic needs controlling the design and construction of stable buildings may be enumerated as follows:

1. Dryness.
2. Warmth.
3. Adequate ventilation but with freedom from draughts.
4. Good drainage.
5. Good lighting, both daylight and artificial.
6. Adequate and suitable water supply.

It is now proposed to consider the outline planning of stable buildings bearing in mind the principles enumerated above.

Siting of stables

To make adequate allowance for the main essentials, great care must be taken in the siting of the stables. In many cases the conditions of the site will force the architect to place the stables in an unsuitable position. However with careful forethought and planning, the deficiencies of aspect and siting may be overcome to a great extent by a full appreciation of the ideal conditions required.

Consideration must first be given to the ground upon which the stables are to be constructed. Ideally the ground should be naturally well drained, i.e. chalk or gravel, and should drain away from the buildings. Clay, which retains water is unsuitable and if stables have to be erected on such land the water holding capacity of the site must be broken down and the water dispersed by drainage.

The buildings containing the horses must be protected from northerly or easterly winds. The doors and windows of the boxes or stalls should therefore face in a southerly direction. On a confined site this aspect may not be obtainable in which case protection must be afforded by other buildings or a belt of suitable trees.

Protection should also be made against the prevailing wind of the area blowing directly into doors and windows. Consideration of the prevailing wind, however, must be related to each individual site, as the contours of the landscape surrounding the site and the relationship of the site to woods, buildings, etc., will have a direct bearing of the affect of the wind upon that site. Therefore get to know the site conditions and plan accordingly.

Although protection from the winds is necessary there should be a free circulation of air around the stables so a site hemmed in by trees and buildings is quite unsuitable. Avoid siting the buildings on top of a hill or in any other very exposed position. To the opposite extreme, avoid hollows which catch the water and are invariably frost pockets during the winter months.

Stable buildings should be positioned well away from adjoining houses and there is little doubt that most local authorities will insist on this requirement as a condition of their consent.

However, such a condition will not necessarily apply in respect of proximity to the owner's house. Many clients will require their stables built close to their own houses not only from convenience, but from an economic standpoint, to reduce the lengths of service roads, drainage and other services, see Layout No. 1. Ease of access to the stables is usually an important consideration with owners who look after their own animals, with only daily help. This will be appreciated by anyone who has had a sick animal in stables requiring frequent visits during the day and night. Care must be taken, in these circumstances, to ensure that the prevailing wind does not carry the smells of the stable into any part of the house. Even the most enthusiastic horse owner will object to the smells of his stable yard being carried into his drawing-room.

Consult the local authority before commencing the design to ascertain any conditions that might be imposed on the buildings as such conditions may materially affect the planning of the scheme. Apart from design requirements it is wise to consult with the authorities and boards on the matter of drains and services, etc., at an early stage.

Layout and the main requirements of the stable group

The units making up the whole of the stable buildings will vary not only in relationship to the size of the establishment but also to the needs on the owner. The requirements for a stable to accommodate say twenty horses without any existing ancillary buildings and allowing for a staff of five grooms living out or at least separately accommodated will be as follows:

1. 20 loose boxes.
*2. 1 sick box.
3. Feed room.
4. Hay store.
5. Straw store or storage for alternative litter.
6. Feed store.
*7. Washing and cleaning room, incorporating drying facilities.
8. Saddle and bridle room (tack room).
9. Utility box or boxes depending on the organisation of the stable.
*10. Litter drying shed.
11. Manure bunkers.
*12. Office, in some cases only.
13. Lavatory accommodation.
*14. Sitting-room for grooms.
*15. Garage or covered area for motor horse box and/or trailer.

In many cases this accommodation will be reduced by the omission of those items marked *, particularly in small establishments.

The relationship of the various units must be carefully considered both in respect of relationship to each other and to the site, and surrounding buildings. A groom should be allowed to concentrate his or, more usually nowadays, her energies on the horses and not be unnecessarily taxed by the necessity to carry bales of hay and straw and sacks of feed great distances. Even in these days of mechanisation there are few stables, particularly small ones, which can afford to install expensive handling equipment. Care must therefore be taken to minimise the handling of heavy materials and a lot can be done in this respect by careful planning. For the uninitiated it would be wise to study the routine of a stable of similar type and size to the one under consideration and make a careful drawing showing the journeys made by one groom during a day. It has already been mentioned that stable routine will vary depending on the type and size and running of the establishment. The routine will also vary between the seasons and the plan must take all eventualities into account to deal successfully with the problem. Remember that a lot of work will be carried out during the hours of darkness. In a hunting establishment most of the daily work will be carried out during early morning while preparing the horses and again during late afternoon and early evening after the horses return. In a riding school similar conditions often occur for horses need to be prepared for a first ride at 09.00 hrs. and the final work cannot be done until the last ride returns in the evening, often just before dark.

To give the designer some idea of the routine in a stable, take a typical day during the season in a hunting establishment. To keep the illustration as simple as possible one horse and one groom has been used as a standard. Remember, however, that one groom may have to look after three and sometimes four horses. The routine in the main is as follows:

Brief routine

COMMENCE 05.30

Enter stable and refill water bucket.

Prepare and feed horse.

Pick up soiled litter and droppings and deposit in bunker.

Fork remaining litter (if of straw) in corners.

Sweep floor and remove dirt to barrow.

After horse has finished feeding, remove clothing.

Hang up clothing to air.

Groom horse and plait mane, etc.

Rug up horse and go to breakfast.

After breakfast remove rugs, tack up horse and re-rug.

The horse will then be taken to the meet, either ridden on or boxed.

Allowing say an hour to get to the meet the horse will leave at about 09.45 hrs. and will probably return to the stables at about 17.00 hrs.

In the meantime:

The groom picks up fresh droppings and takes them to the manure bunkers with the other dirt earlier removed.

Hangs up rugs to dry and air.

Removes retained clean litter to drying shed to dry. When dry, return to box, collect fresh bales and lay bed.

Change water.

Prepare feed and mashes.

Fill hay net or rack.

Generally clean and tidy buildings and sweep stable yard.

ON RETURN, SAY 17.00 HRS.

Give mashes and then the feed.

When the horse has drunk and eaten, give hay net.

Clean horse and remove plaits.

Carry out any treatment to cuts, etc.

Blanket up and bandage legs.

Clean out and refill water bucket.

Clean tack.

LATE EVENING, SAY 21.00 HRS.

Check that horse is comfortable.

Refill bucket.

Refill hay net.

Lock up stable and switch out lights.

This routine is of the simplest and does not take into account the many vicissitudes which may arise and cause increased work. It should, however, by now, be appreciated that the number of journeys and operations in a groom's day can be simplified and made less arduous by careful planning, both of the buildings and the relationship of those buildings to each other.

THE STABLE BUILDINGS

Planning the units considered separately

Having given the architect an insight into the requirements of the siting of the stable group it is now intended to discuss in outline the needs of the individual units. There are two alternatives which suggest themselves at this stage, firstly, that each unit should be explained in full detail or secondly that sufficient information should be given to enable the sketch designs to be prepared, with the fuller details given later in the book. The second of these alternatives has been chosen in spite of the fact that it may be necessary sometimes to refer to the detailed information given in Part 2 while preparing the sketches. Although it is recommended that the whole book be read before a start is made on the designs, it is felt that it will be easier for a person who knows nothing of horses or their management to digest the main essentials of each unit, if they are kept free from the many constructional and furnishing details necessary for their completion. Sufficient material in this part will generally be found to have been given to complete the outline sketch plans, which is the first consideration. The full details of each unit will be given in Part 2, which will cover all requirements for detailed sketches and working drawings. Anybody wishing to digest the whole in detail will find it a simple matter to read Parts 1 and 2 together.

Layout of the loose boxes and stalls

First consideration must be given to the layout of the loose boxes and stalls but before proceeding with the design, the detailed planning requirements of the client must first be ascertained. Most clients will require loose boxes for both horses and ponies but some may want stalls or a proportion of each. Stalls are seldom used nowadays and many existing buildings containing stalls are being converted to loose box accommodation. Stalls have many disadvantages to loose boxes, their only advantages being in economy of construction and labour.

For design requirements, 12 ft.× 12 ft. may be taken as a suitable size for each loose box. This size will comfortably house a 16.0–17.0 hh. hunter. Ponies may require a lesser size but it is felt unwise to reduce a box below 10 ft. ×10 ft., a pony's box today may need to serve for a horse tomorrow. In certain areas of the country and in some specialist establishments where only ponies, particularly the smaller moorland breeds are intended to be kept, the size may be adjusted and then a suitable size should be agreed with the client.

Stalls should be a minimum of 6 ft. wide and 9–10 ft. long.

The basic layout of the buildings containing the boxes will depend on economic considerations as well as on site conditions. It is probably true to say that most stables erected at the present time consist of a simple line of boxes, each opening directly into the open air thus:

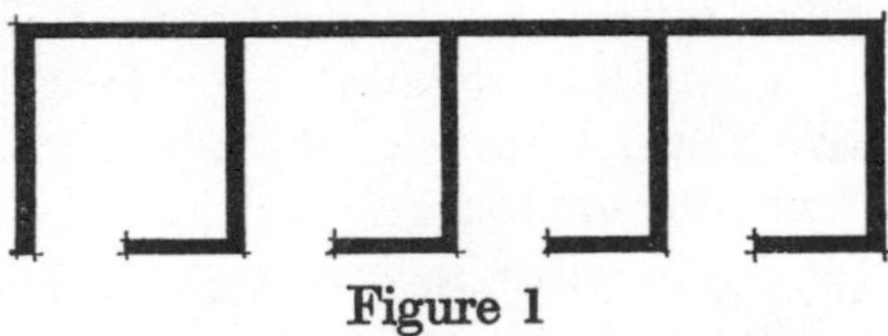

Figure 1

This is the simplest form of layout and quite practical. Depending on the number of boxes needed they may be extended around a yard or a number of yards. The detail can be improved by extending the roof to give an overhang along the open side. This gives a covered walk and helps to protect the horses and grooms from rain and excessive sunlight thus:

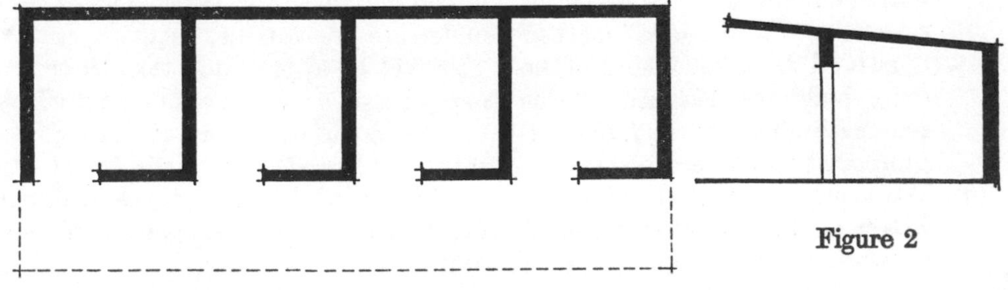

Figure 2

It will thus be possible to carry out most of the work affecting the boxes under cover from the elements. This covered way should be extended to connect to the adjoining buildings of the group. Avoid supporting posts as these can form a hazard.

A more expensive layout is to have the boxes inside a building with the adjacent passage contained within the curtilage of the external walls thus:

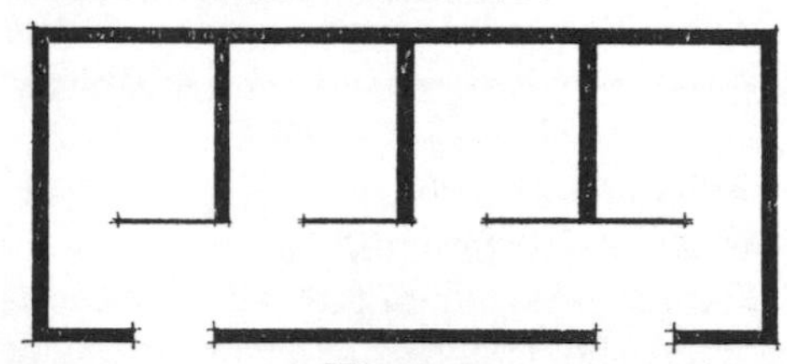

Figure 3

This layout is often found in the more elaborate layouts built during the last century. Ancillary buildings should be built so that they connect to the stables internally throughout the whole range, subject to the conditions imposed by fire precautions detailed later in this book. The advantages of designing in this manner are that:

(*a*) The ventilation of the building may be better controlled and draughts thus reduced to a minimum. A suitable form of mechanical ventilation may be installed as discussed in detail in Part 2. It should be appreciated that ventilation of such an enclosed building will rely entirely on the system installed. If a mechanical system cannot be afforded an adequate cubic content of air must be allowed in the design to offset the "hit and miss" method of natural ventilation. Most of the older buildings designed in this manner indicated that this requirement was appreciated.

(*b*) Work may be carried out under better conditions and with greater comfort.

(*c*) Easier maintenance of warmth.

(*d*) Quieter than open stabling which can be an important factor where stables are built near to main roads, railways or in other areas with a high noise level.

Where site conditions are restricted the boxes may be double banked thus:

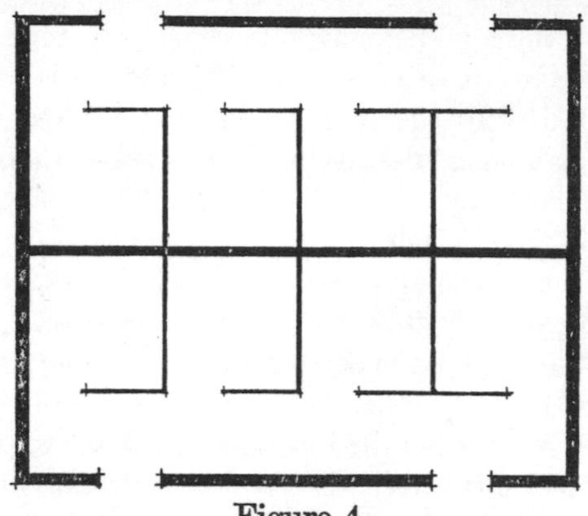

Figure 4

or more economically with a central passage thus:

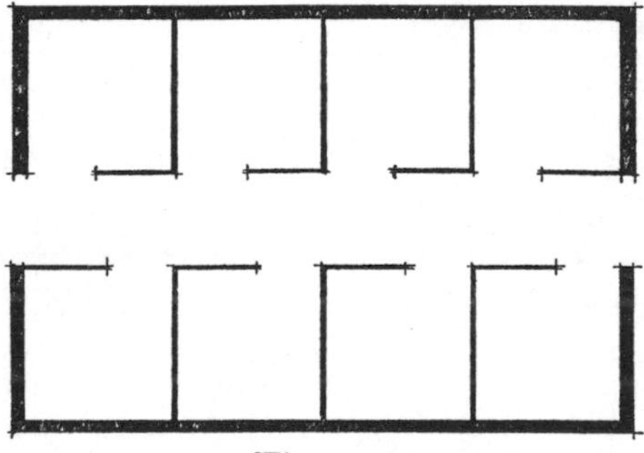

Figure 5

In designs of this layout, where boxes are positioned on each side of a central passage, certain additional precautions must be taken which will be referred to in the detailed planning in Part 2 and under fire precautions.

It will be appreciated that there are many possible variations in layout which can be incorporated in the design to satisfy the needs of the client and site conditions, whilst at all times conforming to the basic requirements of each unit.

To further assist the designer on the subject of layout and to give him a completed picture of the relationship of the units within the stable group three completed layouts are illustrated at the end of Part 1. The client's instructions for and the site conditions of each scheme have been included and discussed.

Utility box

In all stables, of whatever size, it is an advantage to provide, for the want of a better name, a "utility" box. Such a box may be used for a variety of purposes, clipping, grooming, shoeing, washing, and the treatment of cuts, etc. Small establishments of the type illustrated in Layout 1 may need only one, larger establishments more. Layout 2 shows an arrangement which may be used with advantage in all establishments, immaterial of size. In this establishment the management was based on one groom to three horses and a utility box has been provided for the use of each groom. In this box all the various activities mentioned above, are carried out with greater efficiency and cleanliness.

The box, on the basis shown (ratio 1:3), need not be larger than the normal loose box. When, however, it is provided for a greater ratio it should be related to the number of horses likely to be accommodated within it at one time. The eventual size must therefore be discussed and agreed with the client. Its position should be as near as possible central to the group of loose boxes it serves.

Sick box

At least one sick box is essential in large establishments and it is obviously of great advantage to include for one in the plans of even small stables. In the main it is intended for the accommodation of an animal suffering from an infectious disease and therefore requiring to be isolated from other horses. It may in fact be used by any horse needing quiet and possibly specialised treatment.

The box should be placed well away from the stables but as a sick animal requires to be visited more often than a healthy one the position must be related to the convenience of those in charge of him. If possible place it in such a position that although isolated the horse can see the other horses. Remember that horses are gregarious animals, so the patient will be happier and probably make a quicker recovery if he does not feel completely isolated from the world.

The box should be bigger than the usual box by about 50%. The fitting of the box will be fully discussed in Part 2 but it would be wise at this planning stage to remember that the roof must be made sufficiently strong to support a sling attachment.

Feed room

The feed room is intended to house the bins containing the feed for daily use, as opposed to the feed store which will contain the sacks of food or the containers if bulk storage is used. In small establishments it might combine with the feed store or in fact be formed as a recess out of the stable building, though this minimum arrangement is not recommended. The size will depend not only on the number of horses it serves but also on the client's arrangements for buying feed. A farmer might grow his own and for instance send sufficient for a week's supply across to the room from his main store, after carrying out any treatment needed. Some clients who buy from merchants will have regular weekly or fortnightly deliveries.

It is in this room that the feeds are prepared each day and it must be positioned close to and preferably directly connected with the loose boxes.

The equipment to be contained in the feed room consists of separate bins for oats, bran, barley, nuts, chaff, etc. (the bin requirements will depend on the method of feeding the horses and will often vary from one establishment to another), a sink provided with hot and cold water, a bucket filling tap to each service, and wall racks on which to hang sieves, measures, brooms and buckets. Few small establishments will install machinery for bruising oats, chaff cutting, etc., but large establishments may require these machines, and if so allowance must be made for them in the design. Most merchants nowadays will carry out any processing required before delivery, so the need for machinery will be rare.

Some clients will have serviceable equipment and machinery which they will require to be incorporated in the new buildings; obtain the sizes of such equipment or if new equipment is to be bought, obtain the necessary details from the manufacturer. Any machinery to be installed must be studied whilst in operation to ensure that adequate space is apportioned for it. This must be done to provide an adequate safety factor in respect of both the operator and other persons in the room, and also to ensure ease of operating the machine. If any doubts exist on the area to be allocated, design on a generous scale to avoid the possibility of having cramped, and therefore dangerous conditions after installation.

Feed store

This store should open directly into the feed room and the storage areas for hay and straw should adjoin it. The necessity for this room will depend on the supply factors mentioned in the last paragraph dealing with the feed room. Many small establishments will not need a separate store and will combine the feed storage with their hay and straw storage.

The feed in this store will most frequently be contained in sacks, the weights of which will vary from 40 lbs. to 1 cwt. Some large establishments will have their feed in bulk containers. In the cases where sacks are used these will not only have to be unloaded from the lorries and stacked in the store but will have to be taken to the feed room for filling the bins. It may be necessary therefore in some cases to consider the needs for hoists and carting arrangements.

One possible arrangement, where site conditions permit, is to position this store over the feed room, thus:

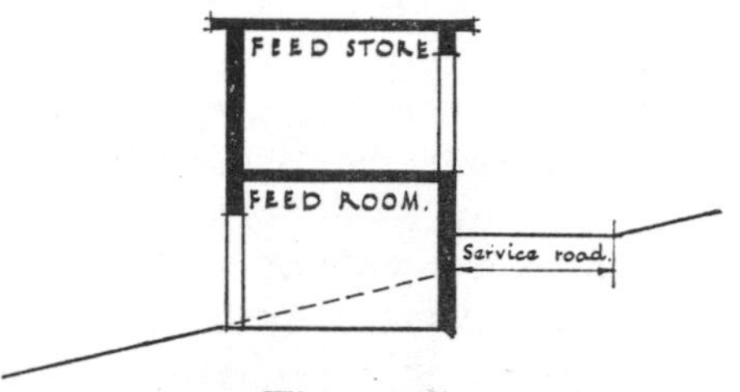

Figure 5A

Ensure, however, that if the feed room is next to the loose boxes that the noxious fumes from the boxes cannot affect and damage the feed. The reader is here referred to the chapter on fire precautions in respect of this layout.

The loading bay indicated in the above sketch allows the sacks to be easily removed from the lorry by means of either a hand or mechanically operated trolley or by an overhead gantry, again either hand or mechanically operated. The feed from the store may be fed to the bins or machinery, dependent on individual requirements, by means of chutes either adjustable or fixed.

Large establishments may have their feed delivered in bulk in which case suitable sized hoppers must be incorporated in the design with adjustable chutes or feed pipes projecting into the feed room and operated and controlled from this room. This type of installation will require specialist advice and construction so it will be necessary to discuss this part of the planning in detail with both the client and the manufacturer before sketch plans are commenced. Many merchants now possess tankers for the delivery of grain for bulk storage. The grain may be pumped to high level from the tanker, directly into the hopper. The quantities some merchants are prepared to deliver by this method may be found to vary from place to place, so it will be wise to ascertain the local conditions before finalising the design.

In most establishments the store will be placed at the same level as the feed room in which case mechanical handling if required can be incorporated by a simple overhead gantry either hand or power operated. Conveyor belts might be used in some cases but as they would be required at low level they would be somewhat of a liability and at the same time will take up valuable floor space. The overhead gantry is probably the most economical and safe provision to make. If installed it should be extended to serve the hay and straw storage areas at least; extending it to other sections of the stable group will in all cases be advantageous as the man-handling of heavy materials will thereby be reduced to a minimum.

Where not possible to install mechanical equipment the sacks, and as later discussed the bales, will need to be carried on trolleys, so ensure that no steps are formed between any two compartments. If a change of level is unavoidable form a ramp and make it as gentle as possible. Ensure that doors are of adequate width, a minimum of 3 ft. 6 in. clear opening is recommended, though this may be required to be increased to a minimum of 5 ft. 0 in. if the use of a mechanical trolley is envisaged.

Hay and straw stores

These stores have been combined within this part, as the requirements in respect of their basic design are identical. Much that has been written above on feed stores, will equally apply to the hay and straw storage buildings, particularly in respect of handling considerations.

The stores should be situated next to the feed store, with easy access to the feed room and to the loose boxes. Hay nets will generally be filled in the hay store and then be taken direct to the boxes. If hay racks are fitted, the bales will then be taken direct to the boxes before breaking and dividing into the racks. Straw will be required at the boxes baled, and the bales are then broken before dividing between the boxes and spread to form the beds. Small quantities of hay and sometimes straw will be required in the feed room, but most will be taken direct to the loose boxes.

An overhead gantry taken from these stores through the feed store, the feed room and then into the boxes would be an advantage if it can be afforded. This arrangement would raise complications in respect of the fire precaution recommendations later discussed. If the stores are placed at first floor level, a hoist must be provided to lower the bales to the level of the boxes. Chutes, or a trap door, fitted to discharge the bales directly into the area of the boxes must not be fitted, as this allows direct access between the boxes and the area of highest fire risk in the entire stable group. If a chute is required it should discharge outside the area of the boxes, for instance into a corner of the feed room.

The size required for the stores will depend on the size and organisation of the stables. A small establishment may not require more than say 10 tons each of hay and straw at a time. However, the tendency nowadays is to buy hay and straw, particularly the former, in large quantities to take advantage of the often extreme fluctuations in both quality and price which occur seasonally.

As with the feed store, it may also be found that hay and straw storage buildings will not be needed. A farmer may grow his own and already have storage buildings for each. In the case of hay he may prefer it left in the rick and cut as required. In such cases one small store suitable for about one week's supply of each material will adequately provide the necessary storage for hay, straw and feed.

Details of sizes, weights, etc., are given in Part 2. Ascertain the client's needs and, with the information given, the architect should have no difficulty in arriving at a suitable size for the accommodation required. Remember, however, to allow for about 10% extra storage area over and above the main requirements. The stores are not emptied entirely before the next delivery is sent.

Storage for other forms of material used as litter

So far in this book straw only has been considered for use as litter. There are many other materials used, many being individual to certain parts of the country where local supplies are available. The storage of these alternative materials must be considered.

These main alternatives may be enumerated as follows:

1. Fern leaves.
2. Fir needles.
3. Peat.
4. Sand (not sea sand).
5. Sawdust.
6. Wood shavings.

In some cases combinations of these materials are used, i.e., sawdust and peat.

The main essentials for the storage of any of these materials is the same as discussed for straw. They must be kept in a dry and well ventilated building. Peat is usually stored either in bales or in bulk and sawdust and wood shavings in sacks. Fern leaves and fir needles are generally stacked in bulk and require to be frequently turned.

Saddle and bridle room

This room, now more generally referred to as the tack room, should be positioned close to and preferably directly connected to the loose boxes under cover. In a large layout more than one room may be required (see layout No. 2).

The size will depend on the number of horses it serves and to the purposes for which those horses are used. The needs will vary from one establishment to another but the client should be able to give the necessary information to the architect for his present and possible future needs.

The layout requirements of the client will also affect the size of the room. Some like the tack of each horse grouped together, others prefer to separate their saddles, bridles, girths, etc.

The only satisfactory way to ascertain the correct size of the room is to set out the walls in detail. To assist the architect in doing this, the spacing, heights, etc., of racks and hooks for the main pieces of tack are discussed in Part 2 and shown on pp. 53, 55, 57. It should be appreciated that the dimensions shown allow for full-sized pieces in all cases.

In addition to the storage of tack this room usually accommodates the medicine cabinet and poison cupboard. These are usually shallow cupboards about 6 in. deep, hung to the wall and fitted with either glass or plastic covered shelves to facilitate cleaning. The poison cupboard must be clearly labelled "POISONS" and both should be fitted with strong locks.

In some establishments a bit case may be required, the size will be dependent on the number, types and sizes of the bits to be stored. The normal bit case is a shallow cupboard about 4 in. deep, hung to the wall and fitted with glazed doors.

Chests for clean blankets, sheets and other clean clothing are usually provided in the tack room. These may be formed as built-in chests or standard chests may be obtained and space for them along the walls allowed. Full details in respect of such equipment is given in Part 2.

This description of the requirements of the tack room has assumed throughout that a separate washing and cleaning room will be provided. In most establishments the tack room will be required to serve both purposes, in which case the fittings and services discussed in the next section, which deals with this room, must be incorporated in the tack room.

Washing and cleaning room

It is of considerable advantage to provide this room in any establishment, however small. It should open directly out of the tack room and have an external door off the stable yard so that all dirty tack, etc., may be taken into it without the necessity of passing through the tack room.

It should be fitted with a large and deep sink or sinks, depending on size, and each sink should be provided with a constant supply of hot and cold water. Facilities may be required for washing and drying blankets, sheets and other clothing, though nowadays, apart from bandages and small articles these are sent out to the dry cleaners. A satisfactory method of airing blankets is to provide one or more of the old type kitchen clothes airers which are suspended from the ceiling and fitted with cords and pulleys. This allows the clothing to be hoisted out of the way while airing or drying. Alternatively electric airers may be provided.

Saddle room horses will also be needed. The client may have these, in which case they must be measured and due allowance made for them. If new ones are allowed for, the size required should be agreed with the client and details obtained from the proposed manufacturer. In some cases these horses may be fixed permanently in position but in most cases they will be fitted as removable furniture.

Bridle cleaning holders will also be required. These are suspended in suitable positions from the ceiling and may be of either a fixed length or of a telescopic type. Each must be fixed in such a position that a clear area all round is allowed for cleaning purposes.

The only other items required in this room are suitable cupboards and drawers in which to store the cleaning materials.

Litter drying shelter

After the spoiled litter has been removed from the horses' bedding each morning, the remaining clean litter, fit for re-use must be dried and aired. During fine weather it may be spread outside the boxes in the stable yard to dry in the sun and wind. When raining and during many of the winter months this method of drying is impossible, therefore some form of covered shelter is needed. This building usually takes the form of an open-sided shed set clear of the other buildings on at least two opposite sides to facilitate the free passage of air. It should be positioned convenient to the boxes. Allow the roof to overhang the open sides to protect the litter against driving rain. Moveable slatted screens are sometimes provided for side protection but it is better to leave the sides as open as possible and rely on the roof only for protection. The floor should be raised and formed as a grille to allow the air to circulate below the litter and so accelerate the rate of drying.

Manure bunkers

The storage provision for manure must be positioned well away from the area of the loose boxes but will require to be easily accessible from them. It must be adjacent to a road or drive to facilitate collection.

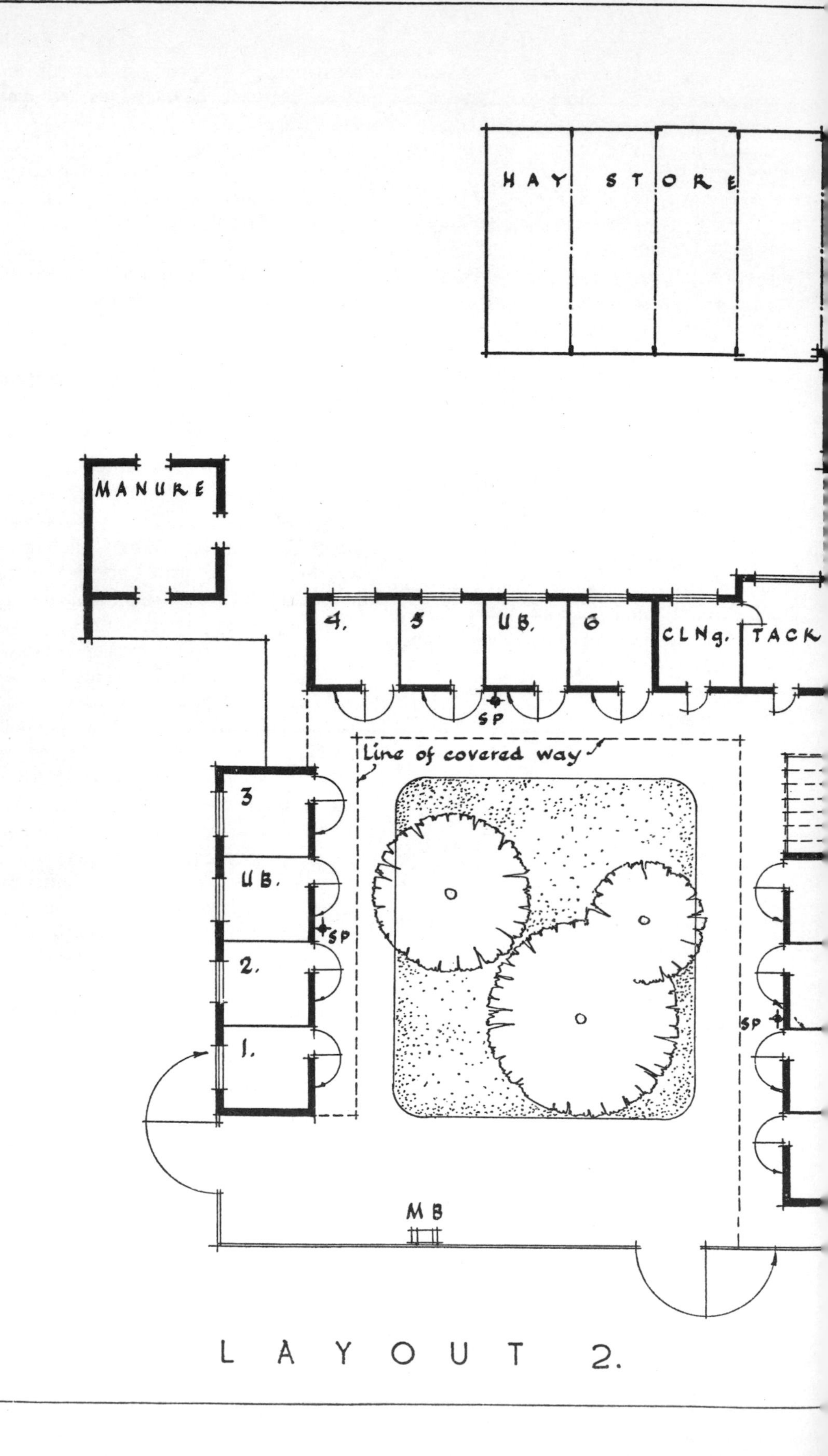
HAY STORE
MANURE
4.
5
UB.
6
CLNg.
TACK
SP
Line of covered way
3
UB.
SP
2.
1.
SP
MB

LAYOUT 2.

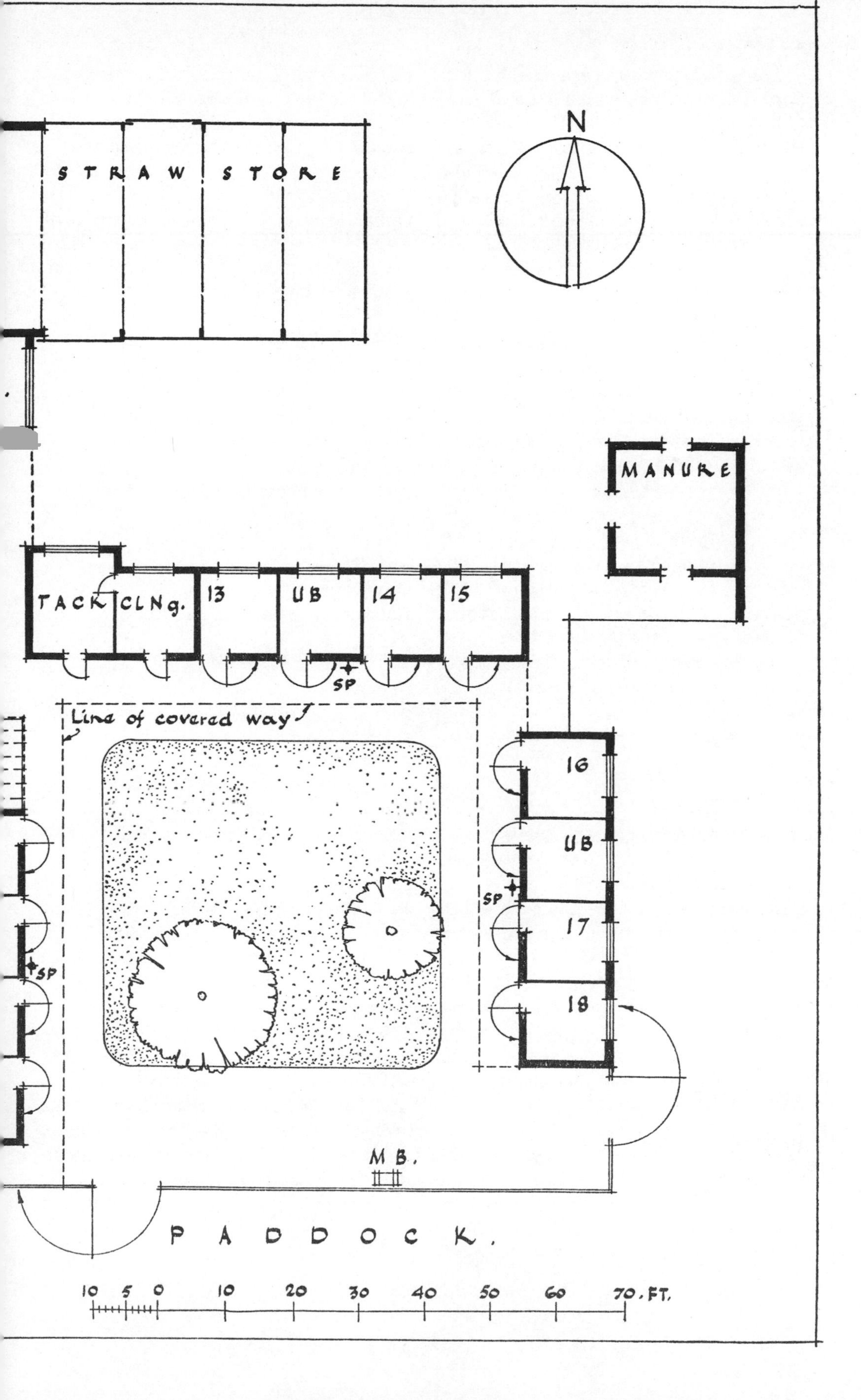
STRAW STORE
N
MANURE
TACK
CLNg.
13
UB
14
15
SP
Line of covered way
16
UB
SP
17
18
SP
M B.
PADDOCK.
10 5 0 10 20 30 40 50 60 70. FT.

The usual formation is of open bunkers and such an arrangement is quite satisfactory. The size and cubic content of the bunkers will depend not only on the number of horses accommodated but also on the routine which each individual stable will follow regarding the disposal of the manure. Some establishments will have contracts for weekly collection or even at lesser intervals. Stables attached to farms will often deposit the manure daily at the permanent stacks or in pits so that it may be allowed to rot down and be used by the farmer on his own land when required.

To arrive at the size of the bunkers to be provided the amount of soiled litter per day must be calculated. This will depend on the use made of the horse and the number of hours per day that it will be out of the stable. Owing to the fast compression weight of manure it is difficult to arrive at any exact figures. Experiments however have been made under varying conditions of weather and heights of stacks and the recommended sizes based on the results of these experiments is given under the detailed planning in Part 2.

Weighing machine

A weighing platform will sometimes be required particularly in large establishments. It should be positioned at the entrance to the stables, preferably next to the office. The indicating dial should be fixed either in the office or be clearly visible from it. Such a machine is a specialist provision and the size and type needed must be agreed with the client. Details of the installation must then be obtained from the manufacturer and provision made accordingly.

Office

Large establishments and most riding schools will require an office for the manager. In most cases it will require to accommodate a desk, chairs, filing cabinets and stationery cupboard. A room of about 100–150 sq. ft. will be ample in most cases. A telephone will be required which should be fitted with external bells.

The office should be positioned to command good supervision over the stable yard and over the delivery of goods. In riding schools supervision over riders passing in and out of the yard must be allowed for.

Mounting block

Most stables will require a mounting block. This should be positioned at the side of the stable yard but should not obstruct the free use of paths or drives. A horse is mounted from the nearside and there should be adequate free space to be able to lead the horse up to it in a straight line and ride off in a straight line after mounting. No horse will stand still to be mounted if its nose or tail is close to an obstruction.

Fences and gates

Fencing and gates adjacent to or surrounding the stable yard and those surrounding paddocks should be of stout construction. The more usual type of fencing is post and rail. Such fencing should preferably be of oak and be of morticed construction, not nailed. Ensure that the main posts are adequately set in the ground (minimum 24 in.) and be concreted in. Gates for lorries should have a clear opening of at least 10 ft. (12 ft. is better) and hand gates should be 4 ft. clear opening to allow the easy passage of horse and rider.

Gates will frequently be used by mounted riders so they should be fitted with hunting latches to facilitate opening. Do not fit gates with closing stiles close to buildings, walls, or return fences, as such positions make it difficult to operate the gates whilst mounted. As far as possible keep all closing stiles to gates one bay (9 ft.) or more clear of all obstructions.

Accommodation for motor boxes and trailers

Most small stables will require accommodation for one trailer and many will own their own motor box. Provision may be required for these vehicles, either by a completely enclosed building, or by an open sided shelter. Normal garage provision should be made, though large establishments with more than one motor box may require a workshop for a mechanic and an inspection pit. In most cases maintenance repairs will be carried out at the local garage.

Full details in respect of sizes are given in Part 2 but ensure that there is adequate space for turning, for lowering ramps both side and rear, and for loading and unloading the horses. As mentioned in the case of the mounting block a horse should be brought up to the ramp in a straight line, and plenty of space should be allowed, both to ensure this and to deal with cases of horses difficult to box.

Staff Accommodation

Few small establishments will require accommodation for staff. Where the stables are close to the client's house facilities are often provided within the house itself.

In large establishments some permanent living-in staff are usually employed, the balance being made up by girls living locally and travelling to and from the stables daily. In these circumstances married staff may be accommodated in cottages or flats, and single staff in bed-sitting rooms or flatlets. Any stable with staff exceeding six, even daily staff, should be provided with a sitting-room, though the tack room more usually serves the purpose.

Lavatory accommodation will be required in most schemes and if the establishment is likely to be used by the public, provision should be made for both sexes. The larger riding schools may in addition require changing and shower rooms for use by their clients.

Residential riding schools should have accommodation for both clients and staff convenient to the stables. Both clients and staff may occupy one building or be separate, depending on the requirements of the client. It is not the intention to discuss the requirements and planning of this accommodation relating to large residential establishments in this book. The requirements relate closely to the planning of hotels and hostels and there are many suitable books of reference on these matters.

SIMPLE LAYOUTS

Introduction

The accommodation and planning requirements so far discussed in outline will cover the needs of most establishments (excluding the specialist types already referred to) that the architect may be called upon to design. The information given should give him sufficient data to prepare his initial sketch layout drawings.

Before proceeding with the detailed planning and construction of the individual units it is proposed to consider layouts based on the information given. It is felt that to illustrate and discuss the planning of the entire stable block at this stage, rather than at the end of the book, will give the architect a better overall picture of the problem, and also assist his appreciation of the details given in Part 2.

Three simple schemes have been selected as typical of the majority of stable groups likely to be built. In an attempt to explain each scheme clearly, the clients' instructions, the site conditions and other relevent factors have been described. These layouts are illustrated as sketches and no detailed fitting is indicated.

Layout Number 1

This layout has been selected as typical of the smaller type of design requirement which might be entrusted to the architect. It is also the type of design which many owners might need and would wish to design themselves, and entrust the work of construction to a local builder. The layout has been kept fairly elastic and could be easily adapted to suit varying site conditions.

The clients in this case were a husband and wife, living in the country with their three children aged 17, 13, and 10 years. Both parents hunted and shared four horses, the eldest daughter owned two horses and the younger ones a pony each. The children were all members of the local pony club in which they took an active part, and all hunted during the school Christmas holidays. The eldest daughter did a certain amount of show jumping and showing during the summer holidays.

The stable management was supervised by the wife, whilst the husband was at business during the day. Two girls from the nearby village came daily to assist.

The site consisted of the house and garden with two paddocks adjoining. The position was fairly open and sloped gently (about 1:50) towards the road. An existing belt of trees along the west perimeter of the lower paddock acted as a windbreak against westerly winds but no such protection existed on the eastern boundary. The sub-soil consisted of about 3 ft. depth of light clay over chalk and drained fairly well, excepting at the lower edge of the paddock adjoining the road. Exceptional weather conditions were necessary however to render this part waterlogged. The upper parts of the paddocks drained well and therefore no land drains were considered necessary.

The needs of the clients in respect of accommodation and approximate position had been carefully considered by them before instructions to the architect were given. A list of the instructions given were as follows:

(*a*) The buildings were to be close to and easily accessible from the house. A fenced-in and paved yard outside the boxes was considered desirable.

(*b*) Eight loose boxes were required (each later agreed at 12 ft. square) and one utility box.

(*c*) Tack room for twelve sets of tack.

(*d*) Cleaning room.

(*e*) Feed room of sufficient size to store one week's supply.

(*f*) Combined hay and straw store of sufficient size to contain 10 tons of each.

(*g*) Manure bunkers, weekly clearance could be arranged with a local contractor.

(*h*) Covered area only for two-horse trailer.

No feed store was asked for as feed in this case could be obtained at short notice, and regular deliveries at weekly intervals were arranged with the local contractor.

The plan is so simple that it is felt that an explanation of it is unnecessary and the architect should have no trouble in following the scheme if he has fully understood the foregoing chapters. This scheme is discussed later in the book where it is used as an example for certain parts of the detailed planning discussions.

Layout Number 2

The requirements for this layout were for an establishment to accommodate eighteen horses with the possibility of future extensions of an indefinite size. The client required accommodation for his own horses and for those kept at livery. The full accommodation as already discussed in Part 1 was required. The staff requirements of this establishment were based on the proportion of one groom to three horses. Each groom has been provided with a utility box, hence the basic grouping in ranges of four. This arrangement is probably ideal but will seldom be afforded. A more common arrangement for this type of layout would be one utility box to each yard, thereby serving three grooms and nine horses. It will be noted that duplication of certain of these units has been made with a view to a simplification of the organisation of the stables, and a reduction in labour.

The site which formed part of the client's farm was fairly level, in fact it was the only level part of the site which was available for building purposes. The land to the south of the buildings fell away to the valley fairly steeply, with the exception of an area of about two acres which was utilised for schooling and jumping paddocks. The subsoil was of a similar nature to that found in Layout 1 and drained well.

The client's house was nearby, but owing to the levels of the ground it was not economical to build close to it. The head groom and three others were accommodated in the client's house and the remainder of the staff consisted of local girls who came daily from their homes. No staff accommodation was required apart from lavatories for both sexes.

To facilitate and speed working conditions a certain amount of mechanisation was envisaged. Provision was therefore made in the planning to allow for the use of mechanised trolleys. The manure storage near the stables was kept to a minimum as it was intended to clear the bunkers regularly and remove the manure to the main stacks on the farm.

As in the case of Layout 1 the plan is simple and it is felt that no detailed description of the planning is necessary. It will be noticed that the whole of the client's requirements have been satisfactorily dealt with, and the recommendations in respect of the various units and their relationship to each other and to the site as already discussed, have been fully covered.

Layout Number 3

This layout has been designed to accommodate twenty-eight horses and to show the type of scheme already discussed in Part 1, where the boxes are constructed within an entirely enclosed building. It has been incorporated not only from the design point of view but also for use as an example when discussing during Part 2 certain detailed planning aspects of this type of building, in particular in respect of fire precautions and ventilation. To save unnecessary duplication of the illustrations the layout is shown opposite page 74 where it is related to the chapter on fire precautions.

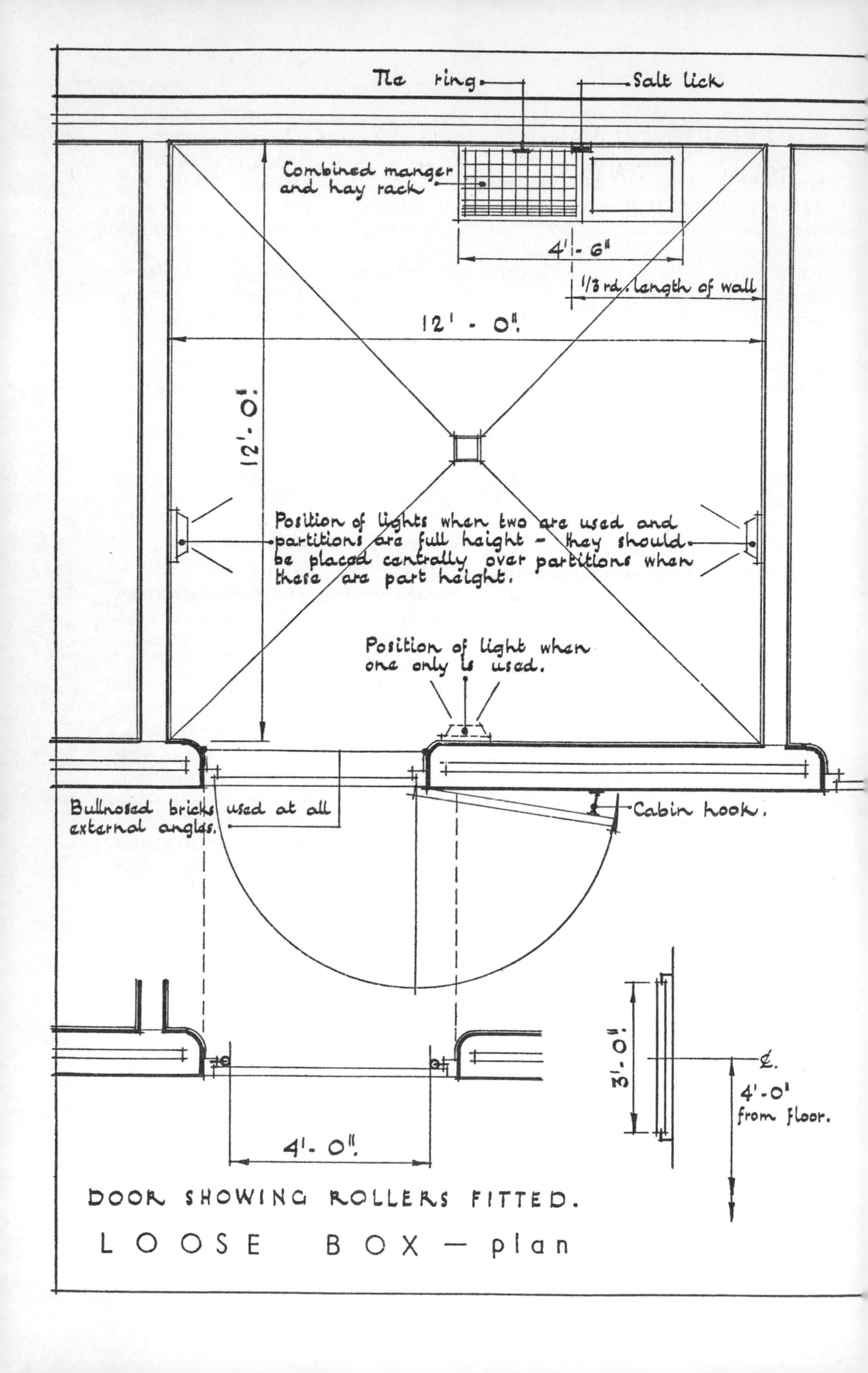

LOOSE BOX — plan

2 DETAILED PLANNING AND STRUCTURAL REQUIREMENTS OF THE BUILDINGS COMPRISING THE STABLE GROUP

General structural considerations

Before commencing the detailed examination of the design and structural requirements of the various components which go to form the stable group, it must be appreciated that no attempt is made to turn this book into a work on building construction. The structural details given relate only to the special constructional needs of each building which may not be fully appreciated by the architect without experience of horse management.

It will doubtless be noted by many architects that most of the materials mentioned both for structure and for finishes are of a traditional nature. These materials have proved themselves to be satisfactory in the positions recommended for many years and must therefore be seriously considered when selecting materials for the structures. It is not intended to imply however that many new materials, or methods of construction now in use, are not satisfactory. Details of the basic requirements of the various parts of each unit are given under the appropriate heading, so when selecting an untried material ensure that its properties are suitable for the position for which it is to be used. If a doubt exists it will be wise to lean towards safety and not use it. Don't try to be different just for the sake of being different and risk failure, good grooms and good horses are difficult to come by and they must not be put in hazard for the sake of fashion or lack of forethought. Apart from the danger factor in connection with unsuitable materials, no client will thank his architect for a floor which breaks up, and becomes saturated with urine in a short time and cannot be cleaned, or for a partition that disintegrates the moment a horse kicks it.

LOOSE BOXES

(See illustrations opposite pages 31 and 32.)

Planning, construction and furnishing

For the requirements of the sketch plan the size of the box was given as 12 ft.×12 ft. This size is adequate, but may be increased up to 16 ft.×16 ft. It need not be increased beyond these dimensions, except in the case of a sick box which will be discussed later under a separate heading. Boxes for ponies may be reduced to 10 ft.×10 ft. Exceptions may be made in respect of some small moorland breeds, though even in these cases this minimum is still recommended to allow the pony adequate space to move around freely. Some clients may favour a box of rectangular shape of say 16 ft.×10 ft., in which case the bedding will be positioned to cover the area well away from the external door. Full details must first be agreed with the client and the design carried out in accordance with his wishes. The uninitiated architect will not in any case be in a position to argue the matter and the initiated will fully appreciate the opposing views, and be forced to accede to the client's wishes even if they are in opposition to his own convictions.

Walls and divisions

Bearing in mind the warmth requirements of the boxes, it is recommended that the external walls should be treated in a similar manner to a domestic dwelling, with a minimum thermal insulation value of 0.30. This requirement will generally entail cavity construction. A further improvement may be made by filling the cavity with a suitable insulation material. For the smaller stable block it will be generally found that load bearing construction using brick, stone or hard concrete blocks is the most economical and satisfactory. Larger buildings may be of reinforced concrete construction, with either infilling panels composed of the above materials, or of in-situ or precast concrete panels. If concrete panels are used, some form of lining will probably be necessary to bring the insulation value up to the required standard. Due to the necessity to stand up to hard usage, including kicking by the horse, all soft or brittle materials must be avoided.

It will be appreciated from the foregoing remarks in respect of the treatment the internal skins of the wall receive, that they may require to be stronger to resist impact than the outer skins. It will probably be found that because of this they will also require to be thicker in proportion. If a 4½ in. brick skin is used it should be reinforced at every fourth course to a height of about 5 ft. Concrete blocks of 6 in. thick or under should be similarly constructed. As a guide it may be taken that all partitioning should be reinforced as stated above if under 6 in. thick. Although there appears to be some evidence that animal urine does not attack mortars made of ordinary Portland cement, if used in properly compacted rich mixes, high alumina cement (in America "Lumnite" cement) is recommended for bedding and jointing brickwork because of the presence of acidic conditions. A satisfactory mix would be 3 or 3½ : 1 sand: cement.

Generally avoid the use of hollow blocks, which although structurally sound under compressive stresses, will not necessarily withstand the aggressive conditions to which they will be subject. Many types of blocks will not provide adequate anchorage for tie rings, used for securing the horse, particularly if it should pull back hard. A horse weighing 1000–1250 lbs. when pulling back can exert a pull on the tie ring of about 1500–1875 lbs. A ring coming free can cause a serious accident.

When it is desired to use hollow blocks, for either economic or thermal insulation reasons, it is recommended that the lower 4–5 ft. of the wall or partition be constructed of solid blocks with the hollow blocks used above this level.

Care must be taken when constructing block partitions to use the correct mix of mortar. A mistake often made is to assume that the mortar mix used for the brickwork will be equally suitable for the block work. The correct mix for blockwork is either 1 : 1 : 6 or 1 : 1 : 9 depending on the time of the year.

Due to the large variety of blocks available for both external and internal construction, it is recommended that information on suitable blocks be first obtained from the Secretary of the Federation of Block Manufacturers. Details of construction, etc., may be obtained in respect of the finally selected blocks either from the above federation or from the manufacturers.

Similar requirements in respect of strength apply to the materials used for internal dividing partitions. These partitions may be constructed of reinforced brick or blocks or by framing, with the main framing built into the walls, and into the floor.

Framing to these partitions may be in steel, reinforced concrete or timber; the latter is not recommended, though if used

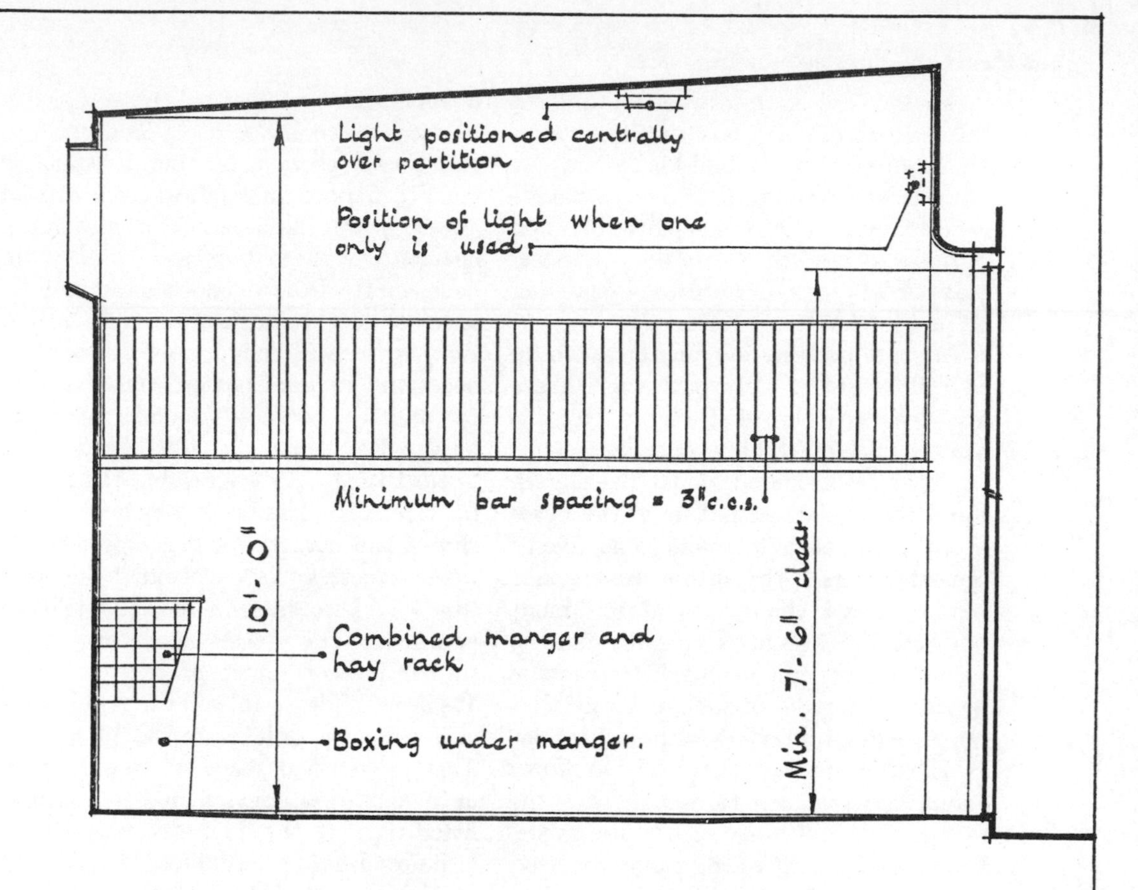

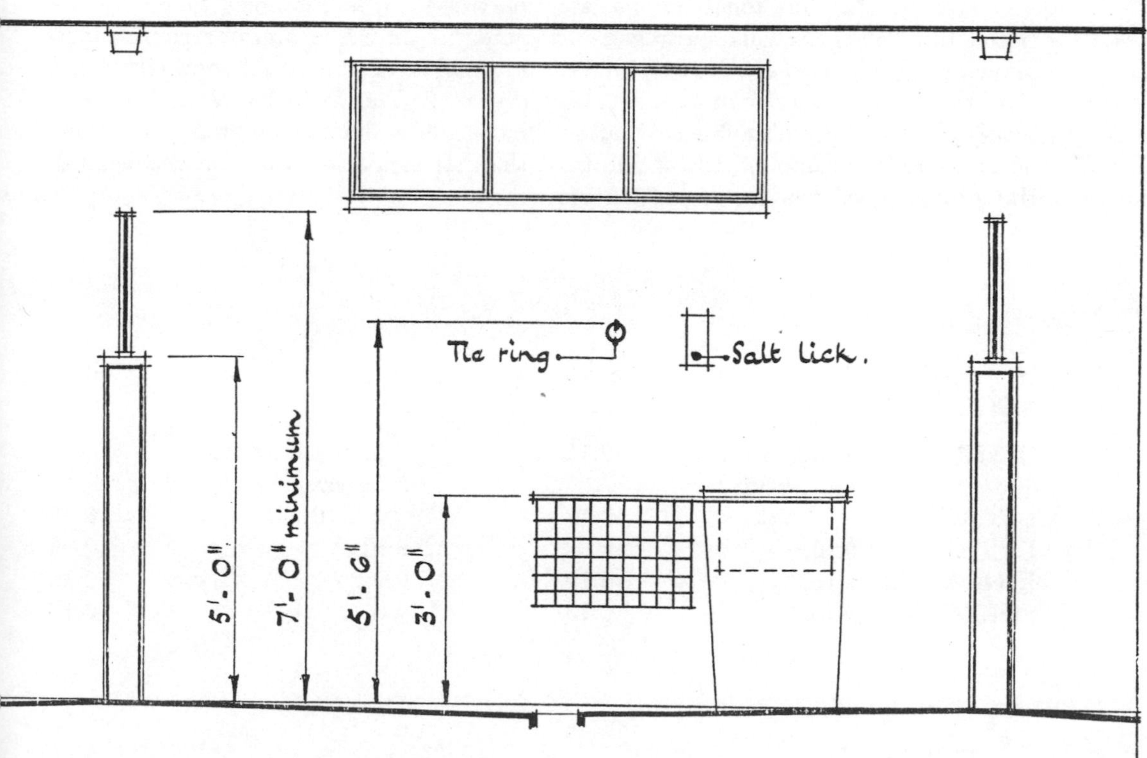

LOOSE BOX — elevations

it should be of hardwood. Standard framed up stall and box partitions may be obtained ready to build in, or may be obtained purpose made to the architect's requirement. The solid parts of such partitions are formed either of steel plate or left open for hardwood panelling to be inserted on the site.

Internal partitioning may be taken up to full height or only part way. There are two schools of thought on this matter both of which have valid reasons for their recommendations. Bearing in mind the gregarious nature of the horse and the fact that it prefers to be able to see other horses, the author prefers that partitioning, in the type of establishment covered in this book, should not be formed for its full height. However the partition must be of such a nature that horses cannot interfere with each other. It is recommended that the partitions should be solid up to 4–5 ft. with the upper section formed of a metal grille. The overall height of the partitions need not exceed 7 ft., but must not be less than this height. This arrangement allows each horse to see its neighbour, but prevents the practice of smelling and biting. It also allows for a free circulation of air within the building, and facilitates the provision of the recommended air changes. This subject will be dealt with in detail later in the book. The grille to the upper section of the divisions is usually formed of vertical steel bars of about $\frac{1}{2}$ in. diameter. The maximum spacing for these bars should be 3 in. centres. Horizontal bars should not be used as these would facilitate crib biting. As this vice will be mentioned on several occasions an explanation would not be out of place. To quote from Mr. Summerhay's Encyclopaedia for Horsemen, "Crib Biting—A disagreeable and harmful vice. . . . The horse lays hold of the rim of the manger, or other projection, with its teeth and sucks wind at the same time." Where internal partitions have panels to the lower part formed of timber it is a sound practice to fix a 12–18 in. high $\times \frac{1}{4}$ in. steel kicking plate each side at floor level for the full length. This prevents damage to the bottom of the partition which in all cases must be fitted tight to the floor.

Before leaving the subject of walls and partitions it is intended to give some guidance on the matter of recommended internal structural linings and structural divisions. The following list of suitable blocks has been prepared and their thermal insulation values noted against each:

LIGHTWEIGHT AGGREGATE		Solid U value		Hollow U value
CLINKER	4 in. × 4 in. with 2 in. cavity	0.18	4 in. × 4 in. with 2 in. cavity	0.17
LECA	ditto	0.13	ditto	0.13
LYTAG	9 in.	0.28	9 in.	0.13
LIGNACITE	9 in.	0.19	9 in.	0.18
SOLITE	9 in.	0.15	9 in.	0.14

DENSE CONCRETE

Walls of British Standard Blocks 18 in. × 19 in.

Single Leaf Walls		
2½ in. thick Solid		0.76
3 in. thick Solid		0.72
3 in. thick Hollow		0.50
4 in. thick Solid		0.67
4 in. thick Hollow		0.48
8¾ in. thick Hollow		0.43
Cavity Walls		
Outer Leaf	Inner Leaf	
4 in. Solid	3 in. Solid	0.35
4 in. Solid	3 in. Hollow	0.29
4 in. Solid	4 in. Solid	0.34
4 in. Solid	4 in. Hollow	0.28
4 in. Hollow	4 in. Hollow	0.24
4 in. Solid	8¾ in. Hollow	0.26
4 in. Hollow	2 in. Solid	0.29

Walls of American Modular Blocks 16 in. × 8 in.

Single Leaf Walls		
4 in. thick Solid		0.67
4 in. thick Hollow		0.50
5⅝ in. thick Solid		0.59
5⅝ in. thick Hollow		0.47
7⅝ in. thick Solid		0.51
7⅝ in. thick Hollow		0.44
Cavity Walls		
Outer Leaf	Inner Leaf	
4 in Solid	4 in. Solid	0.34
4 in. Solid	4 in. Hollow	0.29
4 in. Hollow	4 in. Hollow	0.25
4 in. Solid	6 in. Hollow	0.28
4 in. Solid	8 in. Hollow	0.26

Internal finishes to walls and partitions

The requirements of the internal finishes to walls and partitions are that they should stand up to the aggressive conditions imposed, have lasting qualities, and be easily cleaned and maintained. Materials requiring specialist cleaning treatment or the use of special cleaners or appliances should be avoided. The minimum treatment of a hose down or a wash down with soap and water will, in most stables, be all the treatment they will get, so allowance must be made for this in their selection.

If expense is not the main consideration frost resisting wall tiling, bedded solid either for full height or as a dado of 4–5 ft. high makes a good finish. Alternatively white or salt glazed bricks may be used during the construction of the inner skins of the external walls and the divisions. All external corners should be formed of bullnosed bricks. If a cove skirting can be afforded it will facilitate cleaning.

A granolithic rendering is cheaper and forms a satisfactorily smooth and hard finish. To reduce still further in price a sand and cement rendering will also serve, if used it should be finished with a steel trowel to give as smooth a surface as possible. If no rendering is applied and fair faced brick or blocks are to form the finish, specify that the joints be struck off flush. The reversed struck joint often used with fair faced work will hold the dirt more than a flush joint.

Fair faced brick and block work if left untreated is difficult to keep clean under stable conditions. It will also materially darken the building which should be kept light. Some treatment is therefore called for, the old method of whitewashing is unsatisfactory, distempers do not last and soon look shabby. Some form of good paintwork should therefore be applied and the lower part up to about 5 ft. should be capable of standing up to hard wear and frequent washing.

An economic finish which has proved itself to be satisfactory in stables is "Snowcem" or better still "Sandtex-Matt" which may be applied to either fair faced work or to renderings. If used the lower parts of the walls should be finished by an application of "Blue Circle Clear Glaze" which will allow for frequent washing down.

All hardwood framing and infilling panels will only require an application of oil. There are various seals on the market, but these require regular maintenance to be kept in good condition and it is doubtful if they will stand up to the acidic and humid conditions of the stable. If either seals or varnishes are to be used consult with the manufacturers to ensure that their material will stand up to stable conditions for a reasonable (minimum one year) period, before requiring further treatment.

All grilles and other metalwork should be primed with a good rust resisting primer and may then be painted with oil colour or metallic paint. It would be an advantage to have the whole of the metalwork galvanised after manufacture as this will protect the metal better against the humid conditions.

Roofs and ceilings

The roof may take any suitable form depending on the design requirements of the architect. It is recommended that the combined roof and ceiling should have a thermal insulation value of 0.25. This conforms to the domestic dwelling requirement which was recommended in connection with the walls.

If a double pitched roof is used of 30 degrees or over, the plate line may be brought down to 9 ft. from the internal floor line.

If the pitch is below 30 degrees or a single pitch is used the minimum line of springing should be 10 ft.

Ceilings should not be less than 10 ft. from floor level and in the case of either a low pitch or a flat roof, they will normally be fixed to the underside of the structural members forming the roof. The heights given above may be considered as minimum. Heights may be adjusted above these limits as required by the needs of the ventilation system.

The requirements of the ceiling is that it should be easily kept clean, should be light in colour, should stand up to the acidic and humid conditions of the stable, and in most cases make provision towards the total insulation value required. Hard glossy materials should be avoided as they will allow the moist air to condense on their surfaces.

The old method of render float and set in lime and hair plaster was satisfactory but will not be considered nowadays. Plasterboard, rendered and set, is not nearly so good, hard setting coats should be avoided.

It is fairly safe to suggest that most architects will think in terms of a dry form of construction for ceiling work. When fire resistance is a consideration "Celobestos" is suitable as it will withstand the conditions and has a fire resistance of two hours. When used for fire resisting purposes ensure that the hangers and other supports are also equally fire resisting.

A suitable board for normal requirements is Celotex vapour-check insulation board. This board has been designed and manufactured to insulate under conditions of high humidity and is also unaffected by the acidic conditions of the stable. It is not possible to form a complete vapour seal with any material, some moisture will therefore inevitably penetrate into the air space behind it. Make certain that the roof construction or any additional vapour seal installed behind the vapour check does not inhibit the dispersal of this vapour from the back of the lining.

The following information is given as a guide in connection with the combined roof and ceiling construction using vapour-check board in respect of the thermal insulation value: Corrugated asbestos roof with air space between roof and lining.

½ in Vapour-Check lining only		U value 0.29
ditto	plus 1 in. mineral wool	0.14
ditto	plus 2 in. mineral wool	0.10

1 in. timber, T & G and roofing felt roof with an air space between roof and lining

½ in. Vapour-Check only		0.23
ditto	with 1 in. mineral wool	0.12
ditto	with 2 in. mineral wool	0.09

Floors

The ideal floor to a box or stall from the horse's point of view is that it should be level. However to ensure that drainage requirements are adequately dealt with some slope must be given to it. As the horse may move about freely and change his position in a box, the slope is less important in boxes than in stalls where the horse is tied up and cannot relieve any discomfort felt from a sloping floor. The slope formed, therefore, should be the minimum to allow the urine to flow to its outlet (see chapter on Drainage) and may be taken as a maximum of 1½–2 in. in 10 ft. In the case of stalls the fall should be from front to back.

The additional requirements of stable floors are that they should be dry, smooth, non-slip, non-absorbent and durable. They must be adequately waterproofed if the site is at all damp, and where the natural water level is high, to prevent the heat from the horses body drawing up the damp from the subsoil. If this danger exists a double floor with a layer of asphalt sandwiched between the layers of site and loading concrete should be laid. Unless exceptional conditions apply a reliable bituminous damp proof membrane may be used over the site concrete before the screed is laid.

The surface to many existing stable floors is formed of blue or buff stable bricks or tiles. These cover the requirements already discussed for stable floors but due to their grooved surfaces there is a tendency for them to retain urine and dirt. They are also expensive both to buy and to lay. If used they should be bedded on a 1½–2 in. screed over the damp proof membrane, composed of 3 or 3½ : 1 (sand and cement) and should be pointed with a 2 : 1 mix. When a high alumina cement is used the proportions of the mix should be slightly leaner. Both screed and freshly jointed tiles should be water cured to prevent initial overheating as the material hardens.

Probably the most satisfactory floor and also the most economical is one of concrete, but this must be carefully mixed and laid to satisfy the requirements of the floor. Any concrete floor just won't do, and may quickly break up and become absorbant. The following is an extract from a report prepared by The Cement Marketing Company Ltd., in connection with flooring for stables and adequately covers the requirements for this type of floor.

"The important thing for any concrete, but especially when it is to be used for such purposes as horse box floors, is that it must be dense, well compacted and durable. It must also of course be laid on an adequate and well compacted base of hardcore or similar granular material; the extent of the latter depends on the natural subsoil. Generally the mix should contain at least 600 lbs. of ordinary portland cement to the cubic yard (365 Kg/m^3) and the water cement ratio should not exceed 0.50. With damp aggregates this represents a maximum water addition of about $2\frac{1}{4}$ gallons every hundredweight of cement, or 5.25 litres to 50 Kg of cement, but this will vary with the wetness and the nature of the aggregates. The aggregates (stone and sand, normally 30–40% of the latter) should be sound and well graded material free from contaminating organic matter such as loam and very fine fractions. Water content will be critical and it should be sufficient only to permit adequate compaction with the method available—mechanical compaction (vibration) is really essential. If hand compaction is used then the amount of cement must be increased by 10%.

"If the surface of the concrete is to be exposed to hoofs and general wear a good quality granolithic topping lain monolithically with the base, has much to commend it. Some expertise is necessary in producing a 'grano' topping and the use of specialists for this operation should be recommended. It may not prove possible to provide herring-bone grooves in a granolithic topping and in consequence we think on balance a fine concrete surface incorporating grooves is likely to be the best surface. The use of metal filings or carborundum trowelled into the wearing surface might be advocated also to improve the non-slip qualities."

It will be noticed that herring-bone grooves are mentioned in this extract. In the case of stalls where the gully must be placed at the lowest point of the slope, these may be incorporated, due to the length of the stall and the distance the urine must flow to discharge into the gully.

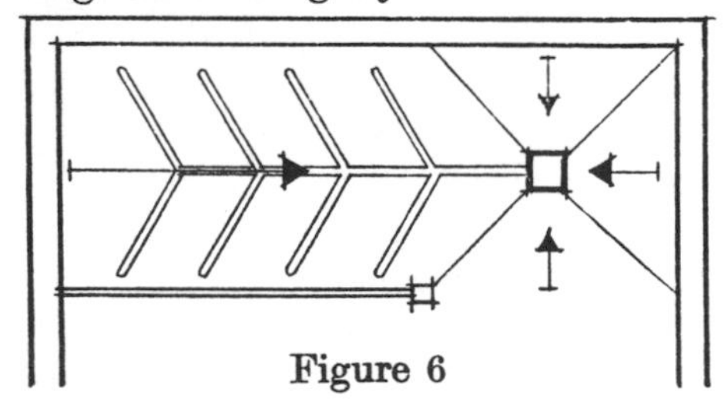
Figure 6

It is, however, doubtful if these grooves do in fact improve the rate of discharge of the urine or adequately control the flow of it. The edges of the grooves, which are their structurally weakest part soon become chipped and the consequent roughness tends to retain a certain amount of the urine and dirt in the grooves. This defect may be partially overcome by rounding all edges. It is considered better practice to retain the smoothness of the floor, to finish the surface with a non-slip treatment and to form the levels adjacent to each stall as shown thus:

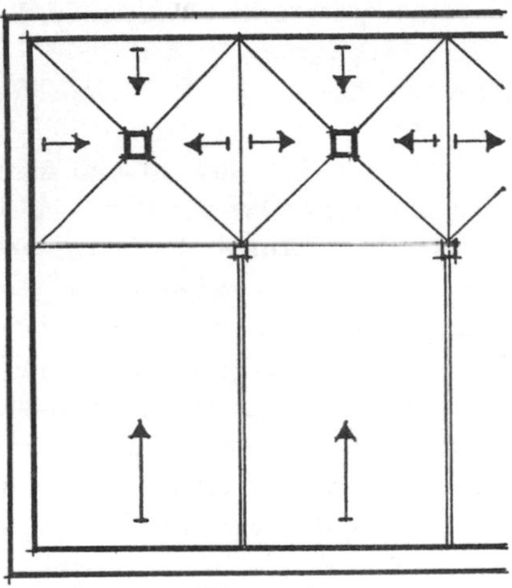
Figure 7

In the case of loose boxes similar recommendations regarding the floor finish are applicable. There are no rules regarding the position of the gully, but from observations it is interesting to note that geldings invariably stale in a position diagonally across the box and towards the centre. Mares also appear to stale as near the centre of the box as possible. There is little doubt that the reason for this is that horses object to splashing themselves, so stale as far as possible to the centre of the litter. On this basis a central position for the gully would be the most satisfactory. The floor of the boxes should therefore slope towards this central position, which due to the shorter length will reduce the extent of slope. This fall should not exceed 2 in. in 10 ft.

One of the dangers in loose boxes is that of a horse getting cast and minimising this danger will be referred to later when dealing with certain fittings in the box. This danger can be overcome by a variation on the design of the box floor. Casting is defined in Mr. Summerhayes Encyclopaedia for Horsemen as follows "Cast: said of a horse when it is lying in a box or stall unable either through lack of space or because it is lying too close to an adjacent wall or division to rise without assistance". The act of getting cast occurs when a horse is rolling and is prevented by an obstruction from completing the roll, the nature of the obstruction at the same time preventing him from obtaining sufficient leverage to roll back to the side at which he commenced the roll.

A method of construction to overcome this danger is to slope the edges of the floor at an angle of 45 degrees as shown thus:

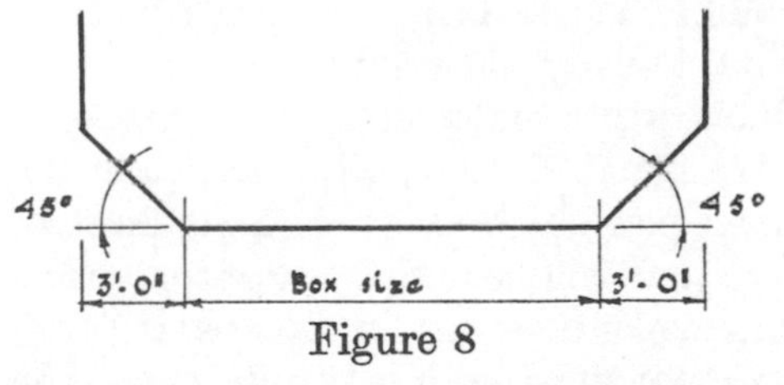

Figure 8

This method eliminates the possibility of a horse getting too close to the walls and therefore will eliminate the danger. The size requirements of the box must, however, be maintained within the sloping edges and will add considerably to the cost of the building.

Doors

Doors to loose boxes should be positioned to one side of the box and in the case of two boxes adjoining they must never be placed next to each other thus:

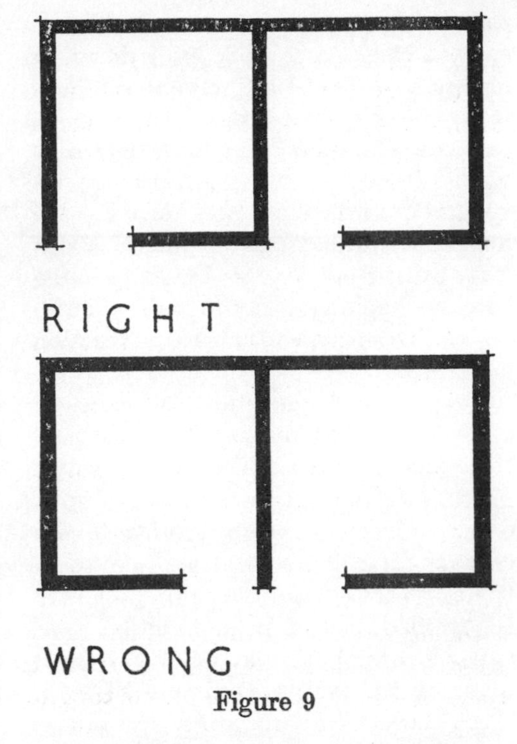

Figure 9

To place the door to the side of the box allows a horse to keep clear of the draught from it, particularly when the upper half is left open.

Doors must be of adequate width to allow a horse to be led out or in without knocking itself. A minimum width of 4 ft. clear opening should be allowed; 5 ft. would be preferable. The overall clear height should make allowance for a horse throwing up its head when passing through and should never be less than 7 ft. 6 in. clear opening.

The door openings should be formed with all external arrises rounded, both to the structural opening and to the frames, no sharp edges should be allowed. An additional protection may be provided by fixing side rollers on each side of the door opening. These rollers may be obtained as standard items of equipment and should be secured to the frame. If used, the clear opening of the door should be measured between the rollers.

Most stable doors are side hung and

must be hung to open outwards. This is essential as a horse while lying down, or in illness or cast may prevent the door from being opened if it is hung to open inwards. External doors to boxes which open directly to the open air are invariably formed in two leaves, the lower section having a height of about 4 ft. 6 in. Both leaves should be hung to close back against the wall when in an open position so that no obstruction is formed.

Doors and frames must be of heavy construction so that they can withstand the rough usage they will receive. Frames should not be less than 4 in.× 3 in. and should be rag bolted to the reveals. The more usual and probably the most satisfactory type of door is of the framed, ledged, braced and boarded type. Framing should be of at least 6 in.×2 in. with a minimum of 1 in. T & G boarded filling, ledges and braces should be of 6 in.×1 in. The junction between the upper and lower sections should be sloped to minimise draughts or a metal strip or angle may be fixed to the bottom edge of the upper half to form a rebate.

A metal bar or sheet metal wrapping should be secured to the top edge of the lower leaf to discourage crib biting. If a bar is used it should be of steel equal in width to the thickness of the door and extend for its full width. Its thickness should be a minimum of $\frac{1}{8}$ in. and it should be securely fixed with countersunk screws. If wrapping is used 1/16 in. steel formed to fit tightly over the top edge of the door is recommended. It should extend for the full width of the door and be secured with countersunk screws. Soft or thin metals must not be used, as a horse might bite through or tear them and thus damage its lips or mouth.

Doors should be hung on heavy steel strap hinges at least 36 in. long for a 4 ft. door. Each leaf should be fitted externally with galvanised stable door bolts, one to the top and one to the bottom of the lower leaf. Some horses have the knack of opening bolts so the lower one to the bottom leaf is essential. A single bolt only may be fitted to the upper leaf about 1/3rd. of the distance from the bottom edge. Anti-kicking bolts and other patent fastenings may be obtained and fixed depending on the wishes of the client. Strong cabin hooks should be fitted, one to each leaf, so that they may be fixed back on the open position when required. The hook to the bottom leaf should be fixed near the top of the leaf.

If required, bars may be provided for fixing across the upper section of the door opening. These are sometimes needed to prevent a horse from jumping out over the lower part of the door when the upper part is left open. These may be obtained from a specialist manufacturer or made to order. The sockets holding the bars should be fitted to the inside face of the door frame.

Some clients may favour the use of sliding doors. The sizes and construction already discussed for side hung doors will equally apply in such cases. Care must be taken when selecting the door gearing. Owing to the amount of dirt that will be swept out of the boxes, channel guides set in the paving will soon get filled with dirt and render the door unserviceable. The most suitable type of gear is one where the door is hung at the head on a roller track and the bottom of the door is guided on an inverted bulb T section.

Sliding doors will require an open panel set in the upper half for ease of inspection from outside the box. Such openings should be covered by a metal grille which should conform to the requirements already discussed in connection with the grilles to the upper sections of the division partitions.

Doors to internal division partitions

are usually designed to match the partitions to which they are attached. They should be hung to open outwards and to fold back against the partition to prevent obstruction to the passage. They should in addition be made to open in line with the main escape route for added safety in case of fire (see chapter on fire precautions). The standard partitioning previously mentioned includes the provision of doors.

Where horses are stabled within an enclosed building (see Layout No. 3) without direct access to the open air from the boxes, the main external doors to such buildings should open outwards and be fitted with an inspection grille or door in the upper panel. They may be formed in two sections or as a single door. For the requirements of doors connecting the various parts of the group the reader is referred to the chapter on fire precautions.

Windows

Windows should where possible be fitted at high level and in the case of an enclosed building where it is intended to install mechanical ventilation they may be fixed lights. However, with the prevalence of power cuts and consequent disruption of a mechanical system, it will probably be better to allow for some opening lights to ensure adequate ventilation during failure of the system. Wherever possible windows should be placed on opposite walls to ensure cross ventilation but not if the windows need to be positioned at low level, as in this position they will allow direct draughts to circulate onto the horse (see chapter on ventilation). Where economy is a serious consideration it is a good plan to form the window as a fanlight over the door.

Where high level opening lights are provided ensure that any fixed gearing to these lights is positioned well away from the horses. If it is found impossible to follow this requirement the gearing must be strongly boxed in so that there are no projecting handles, rods, cords or wires which the horse can bite on or knock itself against.

Low level windows should be avoided but where used they must be adequately protected. They should be strongly constructed, panes should be small and should be glazed with ¼ in. wired glass, In addition a grille formed of steel bars should be securely fixed internally. Similar conditions as those discussed for the upper grilles to the divisions apply to these grilles.

Where windows face onto a passage, they should not project when open, sufficiently either to restrict the width of the passage or form a hazard to a horse passing. Internal fastenings should be so designed that they do not project into the box in whatever position the opening lights of the window happens to be.

To prevent draughts ensure that windows are designed and made to fit properly and that proper provision is made for weathering. Double weathering is recommended in exposed positions.

Stable furniture and services

Each box or stall will require the following items of furniture and services:

1. Manger.
2. Hay rack or ring for hay net.
3. Provision for water.
4. Rings for tying up.
5. Salt lick holder.
6. Electric lighting point.
7. Electric power point if no "utility" box is provided (in the case of stalls one power point may be required to serve up to six stalls).

Manger

The manger is a container for the horse's feed, is usually of metal and should be fixed to the wall of the box at a height of about 3 ft. from the floor. Standard mangers may be obtained constructed of galvanised steel, vitreous enamel, earthenware or stainless steel; they are designed to fit along the face of a wall or in the corner. The latter material is recommended although expensive but the eventual decision will depend on the clients' requirements and economic considerations. Mangers may be obtained combined with either a hay rack or a water trough. Those combined with a water trough are not recommended as they allow a horse to feed and drink at the same time. This practice is not good for the animal's digestion and at the same time usually results in fouling the water (see paragraphs dealing with water troughs and buckets).

One of the objections to a fixed manger is the difficulty in keeping it clean. To overcome this objection and allow for regular and complete washing it could be fitted with a waste pipe. The plug to the waste should be either hinged so that when closed it is flush with the bottom of the manger or be of the remote control type often fitted to basins. An alternative and better method is to have the manger made to fit tightly into a steel framework secured to the wall. This arrangement would allow the manger to be removed to the feed room and properly scoured with hot water at regular intervals.

The usual position for the manger is on the far wall from the door. Before deciding if this position is the best it is proposed to consider the consequence of fixing the manger to alternative walls. In the first place remember that the stables are designed to accommodate horses of varying types and tempers; not only is the comfort and safety of the horse to be considered but also the safety and convenience of the staff.

To facilitate the appreciation of the following paragraphs the walls on the outline sketch of a box have been lettered for ease of reference:

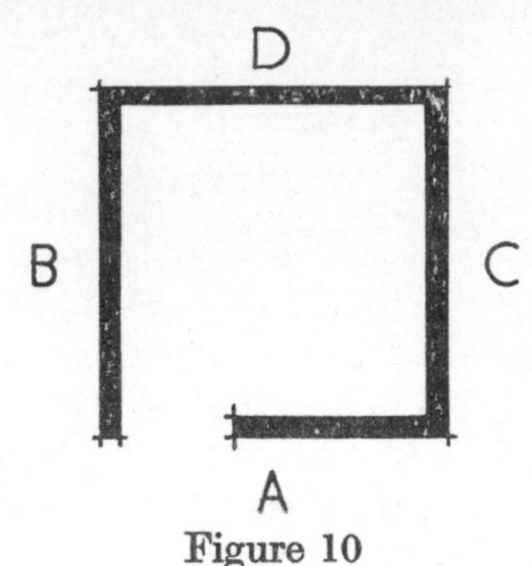

Figure 10

The obvious position from the point of view of convenience for the staff would be to fix the manger on wall A at the side of the door. Consider, however, an ill tempered horse tied in this position. It has only to move sideways when a groom is working on its near side and the door is blocked and so is the escape if required by the groom. A manger placed on wall B would have the same objection.

Walls C or D appear therefore to be the most suitable so far. A horse tied to C but close to A will to a degree give the same objections as mentioned for A and B. If it is tied elsewhere on C, its body will either be in the way of making up bedding or its near side will be too close to D, and will require to be "moved over" frequently. A position between the centre of wall D up to wall B entails the horse being positioned immediately opposite the door. If the top be open he will be subjected to a direct draught which must be avoided. This position also has the disadvantage that anyone entering the box does so immediately behind the horses hind legs.

It would appear therefore that the most suitable position for the manger, and with it as discussed later the hay rack or net, will be about 1/3rd along wall D from its junction with wall C. In this position the safety factor from the groom's point of view is considered and the position of the horse when tied allows a large area of the box to be attended to without disturbance to the animal. The horse is also clear of any direct draught from the door if the top is open.

The space beneath the manger should be filled solid with concrete or boxed in with substantial panelling to prevent a horse injuring itself should it raise a foreleg sharply if startled, or by getting its head underneath it if lying down, and thereby getting cast.

Hay rack

Hay racks may be combined with the manger or may be separate racks which should be securely fixed to the wall. Standard racks may be obtained of suitable sizes and types. The position of the rack and its height should be agreed with the client as there are essentially two schools of thought on this subject. One school maintains that it should be at low level, the horse thereby eating as near as possible in its natural position and the second school prefers the rack at high level, about 5 ft. to the top from floor level. 5 ft. is recommended as a maximum height (*a*) to facilitate filling the rack and (*b*) to prevent dust and seeds getting into the horse's eyes whilst feeding.

Some clients will not require a rack and prefer the use of hay nets which may be tied either to a tie ring or to a patent hay net folder. The position of the net will again depend on the school of thought preferred by the client. Racks or rings for nets, whichever is required should be positioned at the side of the manger. They must not, however, be placed over the water container as this would foul the water.

Provision of water

Horses must be provided with adequate fresh and clean water at all times. The commonest method is to provide it in a bucket. An alternative is to fit a two-gallon automatic drinking trough.

There are a number of valid criticisms both for and against both types of installation. It must be appreciated that a horse will often after and during drinking foul the water by allowing a certain amount to dribble back into the container from its mouth. If the animal has been eating, a certain amount of corn or hay will be contained in this dribble; which if left for a considerable time, may ferment. Fouling of the water by droppings is a frequent occurrence. Having fouled the water some horses will then refuse to drink any more until it has been changed. A horse mainly drinks three times a day, in the morning, after exercise and again in the evening. To ensure a fresh supply, the buckets or troughs must be emptied, cleaned and refilled at least three times per day. It is probably easier to empty and clean out a trough which will then refill on its own accord than to refill a bucket. However, if a stand-pipe is positioned close to each range of boxes (1 to every 6–9) there is not a great deal in it. There is also the consideration that there is more likelihood of the buckets being emptied and refilled than there is of the water trough being regularly emptied and cleaned.

If buckets are used they should not be placed on the floor but be supported on a proper bucket holder to prevent spillage. The holder should be secured to a wall although some owners prefer them fixed to the inside of the door. They must not be fixed close to the manger or below hay nets or racks.

Troughs are obtainable in galvanised steel or stainless steel, and each incorporates its own enclosed ball valve, drain tap and overflow and are of two gallons capacity. The supply service should be in $\frac{1}{2}$ in. tube and the overflow in $\frac{3}{4}$ in. tube. A stop cock, properly protected from interference by the horse should be fitted to the supply pipe within 12 in. of each trough. The space under the trough should be filled in as recommended for mangers.

A stand pipe should be fitted adjacent to each range of boxes and stalls and should preferably be of ¾ in. diameter. It should be fitted with a tap incorporating a hose union.

Ensure that all pipes and fittings on the water supply system are properly insulated against freezing, both those fitted internally and those fitted externally. The recommendations in respect of the design requirements of the water supply system relating to stables are detailed under a separate chapter on water supply.

Rings

Rings will be required for both boxes and stalls for tying up and in some cases for supporting the hay nets. Two rings per box or stall are generally considered to be adequate; if hay racks are used one will often be sufficient.

Rings for tying up should be fixed at about 5 ft. to 5 ft. 6 in. high, one should be positioned close to the hay rack or hay net folder or ring. This will enable the horse to eat its hay while being cleaned or treated. Rings are made for building in and a secure fixing must be ensured. The ring must not pull out if a horse pulls back on it. It would be a wise precaution not to rely upon merely building in but to have the ring bolts made of adequate length to pass right through the wall and be secured on the far side with a 4 in. × 4 in. steel plate washer and bolt.

Rings or folders for hay nets may be fixed at a height agreed with the client. If rings are used they should be secured as recommended above for tying up rings as there will be no guarantee that a horse may not be tied to it at some time. Rings or folders must not be fixed over the water trough or bucket.

Salt lick holder

Many owners like to have a block of salt permanently available for each horse to lick and suitable containers may be obtained for the special blocks which are manufactured for this purpose. The holder should be fitted over or to one side of the manger and at a height of about 5 ft.

Electric points

At least one electric lighting point will be required to each box to enable the duties of the groom to be carried out in a proper manner. It will be appreciated from the outline description of the grooms duties in Part 1 that often the work of grooming and cleaning has to be carried out during dark. Adequate lighting to ensure that the work is efficiently carried out is essential. Two lights one on each side of the box is far better than one. Lights, if suspended, should be high and well clear of the horse at a minimum level of 10 ft. Suspended fittings which for safety reasons require to be at high level are not recommended, they soon get dirty in a stable with consequent lessening in intensity and require regular cleaning which invariably does not get done as often as it should.

It is considered better to use waterproof bulkhead fittings securely fixed to the walls and fitted with either heavy pattern prism glasses or with normal glasses and strong galvanised grilles. These may be placed at about 7–8 ft. high and if heavy quality fittings are selected there is no risk either of damage to the horse, or of the horse damaging the fitting. At the same time being in a more accessible position they are easier to keep clean and bulbs may be more easily changed. Bulbs should be of either 100 or 150 watts each.

In stalls or when box divisions do not extend for the full height one light positioned to shine each side of the division will be adequate.

One power point will be needed to each box, if "Utility" boxes are not provided. This will be mainly used for the clipping machine and in some stables for a grooming machine, and for an inspection lamp. The best position for this point is outside the door of the box about 6 in. above the top of the lower leaf. Where boxes are contained within an enclosed building the point should be positioned outside the partition and adjacent to the door. In stalls one point to six stalls will generally be sufficient.

Switches to lights and to power points should be positioned outside the doors of the boxes though it is better practice to centralise them in the tack room and all equipment must be of a waterproof pattern. The full design requirements of the electrical installation relating to stable buildings is dealt with later in a separate chapter.

Precautions against infestation by flies

A major concern during the summer months in all stables is the control of flies. The use of fly screens over the windows and doors is impracticable. Messrs. Cooper, McDougal and Robertson Ltd., have produced an automatic, electrically-operated aerosol dispenser, called the Coopermatic, which has proved to be successful in stables. This dispenser is powered by a Smith's clock motor which operates a leveler system releasing, every fifteen minutes, a measured dose of a syngerised pyrethrins formulation from an aerosol dispenser. One machine will cope with about 6,000 cubic ft., and operate for eight hours per day. Each dispenser will last about three months.

The use of this type of formulation in loose boxes is of particular advantage as the chief fly pest is commonly the Stable or Biting House Fly (Stomoxyz calcitrans L.). This fly is very susceptible to pyrethrins, being killed by minute doses of the insecticide but, even more important, being strongly repelled by minute doses in the atmosphere. Thus the treated areas do not become littered with dead flies as the flies literally do not enter that area.

Although this is a very convenient method of controlling adult flies in and around the stable areas, nine times out of ten the stable is producing its own fly problems in the manure bunkers. A Coopermatic installation should, therefore, be backed up by a residual insecticide treatment of these bunkers. This can be done with a Malathion Wettable Powder applied with a Knapsack Sprayer every four or five days in the summer season. This builds up a dung/malathion sandwich, thus making it impossible for fly larvae to develop within the material.

It will be appreciated from the above information that one Coopermatic device will serve four boxes, providing the intermediate partitions do not extend to the full height. The device should be positioned about 8–9 ft. from the floor. It is recommended that before deciding on the actual position of the device that the manufacturers be approached and their advice obtained.

STALLS

(See illustration facing page 46)

If stalls are to be provided they should allow for a minimum width of standing of 6 ft. with a minimum length of 9 ft. The passageway at the rear should be a minimum of 6 ft. wide giving an overall internal width to the building of 15 ft.

All the structural and furnishing details already dealt with in connection with loose boxes apply to the construction and fitting of stalls. In addition it is usual to provide pillar reins to the outer ends of the stall divisions. These allow for the horse when saddled and bridled to stand facing the passageway whilst waiting to go out, secured by a rein (usually rope or chain) on each side. These reins are secured with rings to the stall divisions and hooked to the rings of the bit. If pillar reins are required and standard partitioning is used allowance for them may be made with the partition. It should be appreciated that fixed water containers should not be fitted in stalls. Buckets should be used so that the water may be removed while a horse is feeding.

Draw bars, fitted to the rear end of the stall divisions to prevent a horse backing out may also be needed. These can be fitted to the standard partitioning and are either removable or of telescopic type.

UTILITY BOX

The requirements of a Utility box, immaterial of size, are constructionally identical to the loose box.

With regard to the fittings, a manger will not be required although hay racks, or rings for hay nets relative to the number of horses to be accommodated at one time, may, in some cases be fitted. This provision will depend on the wishes of the client. If horses are likely to be washed in this box it will be wise to provide overhead radiant heaters to assist drying. Many owners object to washing horses so this matter must be discussed and agreed for each individual scheme. Some establishments may even require a separate drying room, but this requirement is likely to be extremely rare and such a provision must be discussed and agreed with the client. If such a room is asked for it must lead directly out of the Utility Box.

The box will mostly be used for the remaining uses mentioned in Part 1 and provision may need to be made for a clipping and grooming machine. This machine is usually suspended from a trackway at high level (8 ft. minimum) which is usually fitted across the box a few feet in from the front wall. Details of the machine intended to be used must be obtained from the manufacturers and be fitted and wired in accordance with their instructions. A 13 amp. point will be needed for this machine which may be fitted internally in this case as the horse will be tied up and a groom or grooms will be in attendance whilst the horse is occupying this box.

It would be an advantage to lay on hot and cold water and also increase the amount of lighting discussed in connection with loose boxes. Additional lights at low level, set in the side walls to shine on the lower and under parts of the animal will add considerably to the ease and efficiency of the work likely to be carried out.

SICK BOX

The position of the sick box has already been discussed in Part 1. Most large establishments will require at least one isolated sick box, and the provision of one in even small establishments is advantageous. Where no sick box is provided it would be an advantage to allow for the necessary structural modifications in one box, so that conversion could be carried out at any time without the need to make expensive alterations. Such a box, however, positioned close to other boxes could not be used for infectious cases, which would require isolation.

It is recommended that the sick box, whether isolated or not, should be about 50% bigger than the average box, say 12 ft. × 15 ft. This area will allow adequate room for treatment of the animal.

The main structural variation required is the provision of a beam to support a set of slings. This beam should be fixed at a height of about 10 ft.

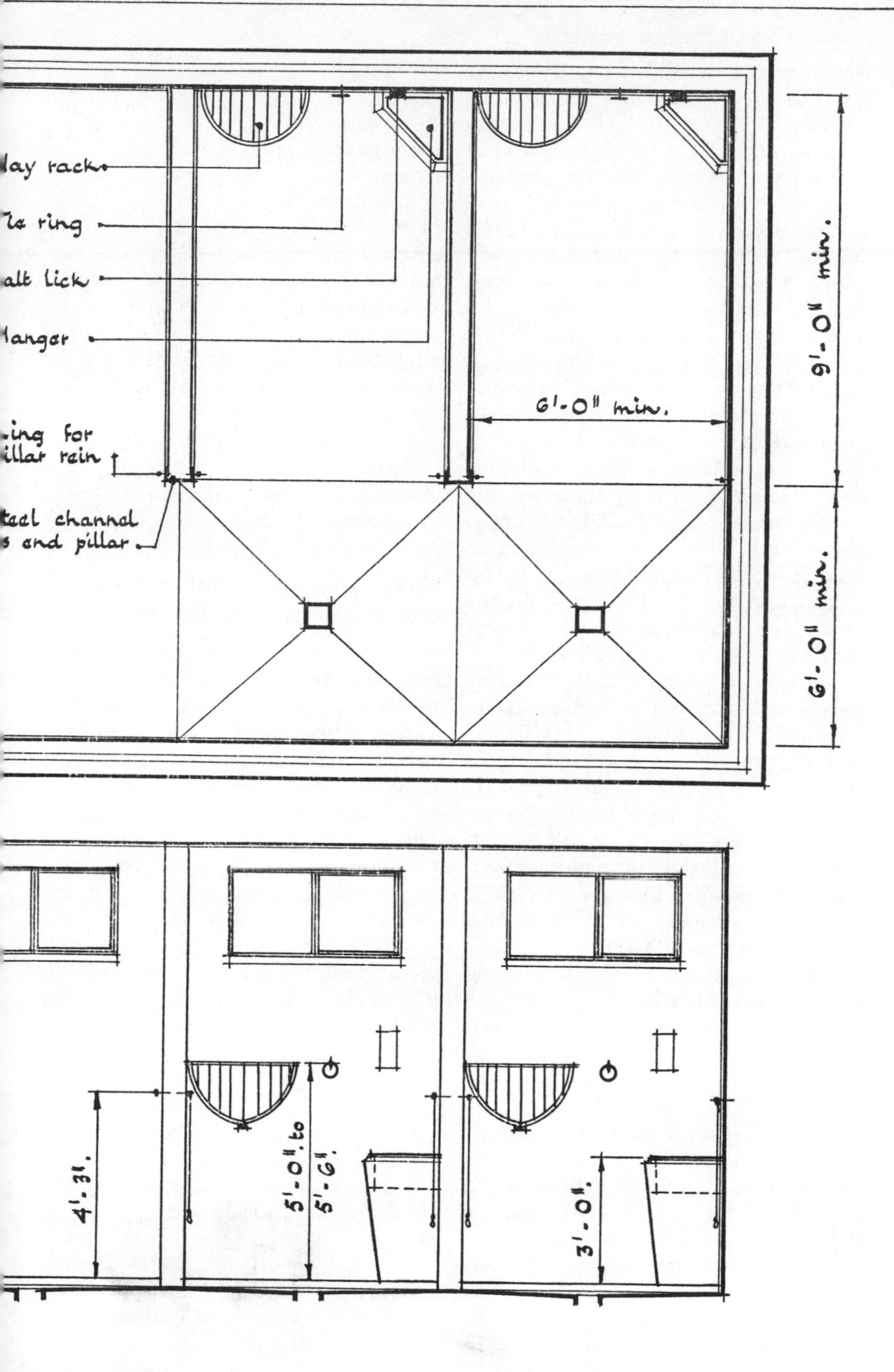

lay rack.
le ring
alt lick
langer
ing for
illar rein
teel channal
s and pillar.
6'-0" min.
9'-0" min.
6'-0" min.
4'.3".
5'-0" to
5'-6".
3'-0".

STALLS

above floor level and must extend for the full width of the box. It must be capable of supporting the full weight of the horse. The design load will depend on the type of horse intended to be kept at the establishment which is the subject of the design. In round figures a heavy draught horse will weigh about 1 ton or a little over. Hunters and hacks of about 15 to 17 hh weigh about 1000 to 1500 lbs. Whilst constructing the box, the difference in cost between a beam capable of taking say 1.25 tons, which may be considered the maximum requirement and one taking, say, 0.75 tons will be negligible, so it would be wise to allow for the maximum load. The load the beam is able to take should be clearly figured upon it. This will save future doubts on its strength and if not designed for the maximum load recommended will prevent the possibility of accidents.

The slings are secured to the beam by means of blocks and an endless chain. The architect will need to notify the manufacturer of the type and size of the beam he intends to install and its height from the floor level. The remainder of the equipment will be supplied and installed by the manufacturer. It is recommended, however, that early consultations should be arranged with the specialist manufacturer so that all details may be agreed before working drawings are started.

A horse when required to be suspended in a sling will require to reach its manger, hay and water from the one position. These must therefore be grouped together to facilitate feeding.

The gregariousness of horses has already been mentioned and that the position of the sick box, if isolated, should face towards either the stable yard or an occupied paddock. Do not therefore defeat the object of this advice by placing the manger, etc., on a blank wall. It is better in this case to place the manger, etc., on the wall containing the door so that the horse may look out if it cares to do so. If the animal requires complete quiet the door can be closed.

Additional tie rings will be required in this box for securing hobbles. They should be fixed at low level about 12 in. above the floor. Two to each side will be sufficient and they should be evenly spaced, one fixed at a point a quarter of a length along each wall from the corner would be suitable. The fixing of these tie rings should be in accordance with the earlier recommendations for fixing.

Apart from these additional provisions the box should be constructed in accordance with the conditions for loose boxes already detailed.

FEED ROOM

(See illustration facing page 48)

The main essentials of the feed room have already been fairly fully discussed in Part 1 and the only outstanding factors in respect of it, are the finishes and the various items of equipment it contains. The illustration shows the layout of the feed room incorporated in Layout No. 1 and is typical of the type of provision likely to be required in a small establishment. It must be appreciated that feeding will not only vary between horses, depending both on the needs of the animal and the work it is doing, but will also vary between establishments. The layout shown cannot be assumed to be suitable for all clients. In all cases the equipment provided should be discussed with the client and an agreement reached at an early stage of the design.

The equipment incorporated in the layout shown includes the following items:

(*a*) Bins to contain oats, barley, bran, nuts and chaff. In other layouts additional bins might be required.

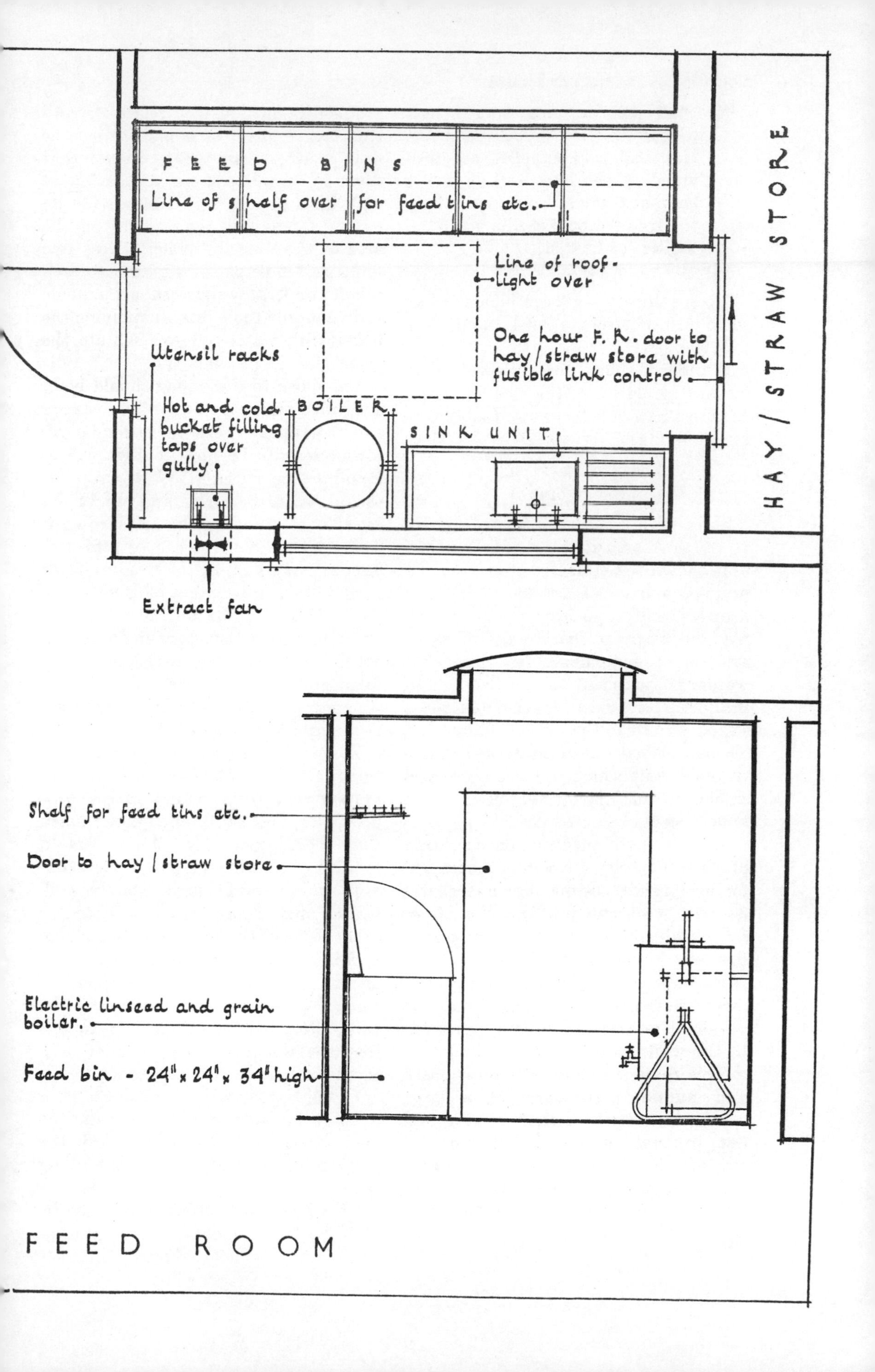
FEED BINS
Line of shelf over for feed tins etc.
Line of roof-light over
HAY / STRAW STORE
One hour F.R. door to hay/straw store with fusible link control
Utensil racks
Hot and cold bucket filling taps over gully
BOILER
SINK UNIT
Extract fan
Shelf for feed tins etc.
Door to hay/straw store
Electric linseed and grain boiler.
Feed bin - 24" x 24" x 34" high

FEED ROOM

(*b*) Sink and draining boards unit with drawer and cupboards under. Hot and cold supplies are provided at the sink and branches have been taken from each supply to serve the bucket filling taps.

(*c*) Boiler for linseed.

(*d*) Racks for equipment.

No machinery has been included but some establishments will need this provision.

Standard bins are shown each measuring 24 in. × 24in. × 34 in. high. Bins of this size will contain the following amounts:

2 cwts. of oats
1½ cwts. of bran
3½ cwts. of cubes.

The sink and double draining board unit shown is a standard stainless steel unit, on a base cabinet containing two drawers and cupboards under, which may be used for the storage of small articles of equipment. Hot and cold water is supplied to the sink and branches have been taken from these supplies to serve the bucket filling taps. These taps have been positioned over a trapped gully which will also serve as a discharge for the water used when floor washing.

A small 3 kW. electric boiler is fitted of ½ cwt. capacity for boiling the linseed for the mashes. The one shown measures 25 in. × 24 in. × 40 in. high. For larger establishments boilers of greater capacities may be obtained.

Racks will be needed for brooms, forks, sieves, etc., and these may be obtained as standard items and fixed to the walls at appropriate heights.

This room is liable to be both dusty and humid, particularly in a large establishment where chaff cutting and oat bruising machinery may be installed. The finishes must stand up to these conditions and be easily cleaned. Although the reduction in the dust problem is outside the architect's control, the problem of humidity may be dealt with satisfactorily. Ensure that there is adequate cross ventilation at high level and in addition provide an extract fan to clear the vapour from the area of the sink and boiler. These two units should be placed adjacent to each other, preferably against an outside wall, not only to facilitate supplying the boiler with water but to facilitate the extraction of the vapour.

The finish to the ceiling should be of a suitable material to withstand vapour and the vapour-check insulating board, earlier described for box ceilings, will be found to be suitable. Walls must be easily washed down and must be capable of standing up to hardwear. Tiling is suitable if it can be afforded, alternatively a finish of Sandtex-Matt with a final application of Blue Circle Clear Glaze for the full height of the walls makes a satisfactory and economical finish. Floors must withstand hardwear and frequent washing down. A concrete floor finished with granolithic as described for the box floors forms an economical and satisfactory finish. If tiling, either tasselated or quarries, can be afforded they give a satisfactory floor with a far better and cleaner finish than the granolithic. There are a number of suitable jointless floors available which would serve equally well. Unless such floors can be afforded in other parts of the buildings and so give a fairly large area to the specialist layers, they tend to be expensive. Lino, PVC, either sheet or tiles, rubber, cork, etc., will not stand up to the frequent washing down required and should not be used in this room.

It will be necessary to wash the floor of this room at least once a day and the floor should be laid to fall towards the gully shown beneath the bucket filling taps.

This room will often require to be

warmed and unless some form of central heating is available a high level radiant heater is recommended. Owing to the humid conditions and the fact that those working here will frequently be using water, the fire must be positioned well clear of both persons and machinery. An electric fire at low level would constitute a considerable source of danger and should not be permitted.

The electrical installation in this room should be a waterproof system throughout as dealt with in the chapter dealing with this installation later in this book. An electric lighting point will be required; more in some cases—particularly in large establishments where machinery is installed which will require to be well lit from the safety point of view. Bulkhead fittings, as recommended for the boxes, are suitable though they need not be of the heavy and protected pattern needed in the boxes. Power points will be required consisting of fused spur units for each piece of equipment. Each should be fitted with a neon pilot light and be engraved to indicate which piece of equipment is served. Starters may be required to some pieces of machinery and these must be fixed in convenient positions and be engraved with the name of the equipment it controls.

FEED STORE

The feed store requires to be a dry, well ventilated and light building. Providing these conditions are complied with and the floor is made adequately strong for the load, the architect is left to his own discretion in respect of it. The description in Part 1 adequately covers the planning requirements and the only outstanding question is that of the capacity. This will depend on the individual requirements of each scheme and will need to be settled with the client at the commencement of the job. A minimum of 10% extra to requirements should be allowed for in the design as fresh deliveries will be made before the old stock is fully used up.

The feed which will be stored in this building will be oats, bran, nuts, barley, etc., and all are invariably delivered in sacks of the following weights:

Oats and barley	1 cwt.
Bran	10 stone
Cubes	70 lbs.

In some cases in the larger establishments feed might be stored in bulk, in which case hoppers will be required. This provision is a specialist supply and details must be discussed with the manufacturers. The building must be designed to suit the size, delivery requirements and the method employed of emptying the hoppers.

HAY AND STRAW STORES

The basic requirements for the storage of both hay and straw are similar, so to save unnecessary repetition the details are combined under one heading. In many cases these two materials will be stored in the same building, particularly in small establishments.

To arrive at a suitable size for the building the following information should be found to be adequate. Both hay and straw are delivered in bales, the size for both being about 36 in.× 18 in.×15 in. They are stacked either flat or on edge. As a rough guide:

1 ton of hay contains 44 bales
1 ton of straw contains 60 bales.

The store must be dry and well ventilated, in many cases an open sided building is used using concrete trusses and an asbestos roof. Such a building may serve the ventilation problem but allows the bales to be soaked by driving rain which should be avoided. A totally enclosed building is the most satisfactory but with adequate cross ventilation provided at all levels.

If brick construction is used the side walls could be formed of open honeycomb brickwork, or louvres could be inserted. No doubt there will be many methods of adequately providing the cross ventilation required which will occur to the architect and which can be incorporated in his design.

The floor, apart from satisfying the loading requirement, will need to be easily kept clean and not dust up. A concrete floor with or without a granolithic finish, laid in accordance with the recommendation for the loose boxes, would be satisfactory. To ensure adequate ventilation around the bales it is best to raise them above the main floor level. This can be arranged by incorporating upstand beams in the floor construction and the bales can then be either stacked on metal or concrete grilles fitted to the top of the beams, or on timber baulks laid across them. By raising the bales clear of the main floor vermin have less chance and are less inclined to get into the stacks. Rats are often a particular nuisance in these stores and if they contaminate the hay it will often be found that the horses refuse to eat it. Adequate space must be left at the ends of these raised parts to allow any loose material to be removed. Additional ventilation to the stack is often provided by the insertion of timbers at intervals during stacking.

Design if possible to allow free space all round the stack and also for about 10% more than the total requirement. Remember that new loads will be delivered before the old is fully used.

Doors to the store for the delivery of the material should be of sufficient size to allow the lorry to back right up to the store, a sliding door of about 10 ft. square will be adequate in most instances for this purpose. If gantries are installed then allowance must be made for them in the design of the doors. The door between this store and any other part of the stable block should be of at least 1 hr. fire resisting construction as this store is main fire risk in the group. This provision applies equally to the separating walls (see chapter on fire precautions).

Electric light will be needed in the store but no other electrical equipment. The bulkhead type fitting is recommended. Ensure that the positions of the lights are not close to the stacking area. A light fitting positioned close to or touching a bale can easily cause a fire.

SADDLE AND BRIDLE ROOM

(See illustrations facing pages 52, 54 and 56.)

Following the main essentials for this room which have been defined in Part 1, the illustrations have been prepared with a view to facilitating the setting out of the racks, etc., in this room. These illustrations show the tack room designed to serve the needs of that shown in Layout No. 1, and allow for twelve sets of tack. The plan and an elevation of each wall are shown and fully dimensioned. By using the dimensions shown and studying the following descriptions the architect should experience little difficulty in setting out a tack room of any size, adapted as required to the needs of his client. In a large tack room space may be saved by using spur walls as well as the main walls for the racks and hooks.

Saddles

These are stored on saddle racks which may be of standard manufacture or specially made, if so desired. The spacing shown at 24 in. centres is adequate for all saddles including side saddles. Ensure that the racks are not placed very close to the floor or the saddles may be damaged by accidental knocking and by floor cleaning. To the other extreme the height of 6 ft. to the top

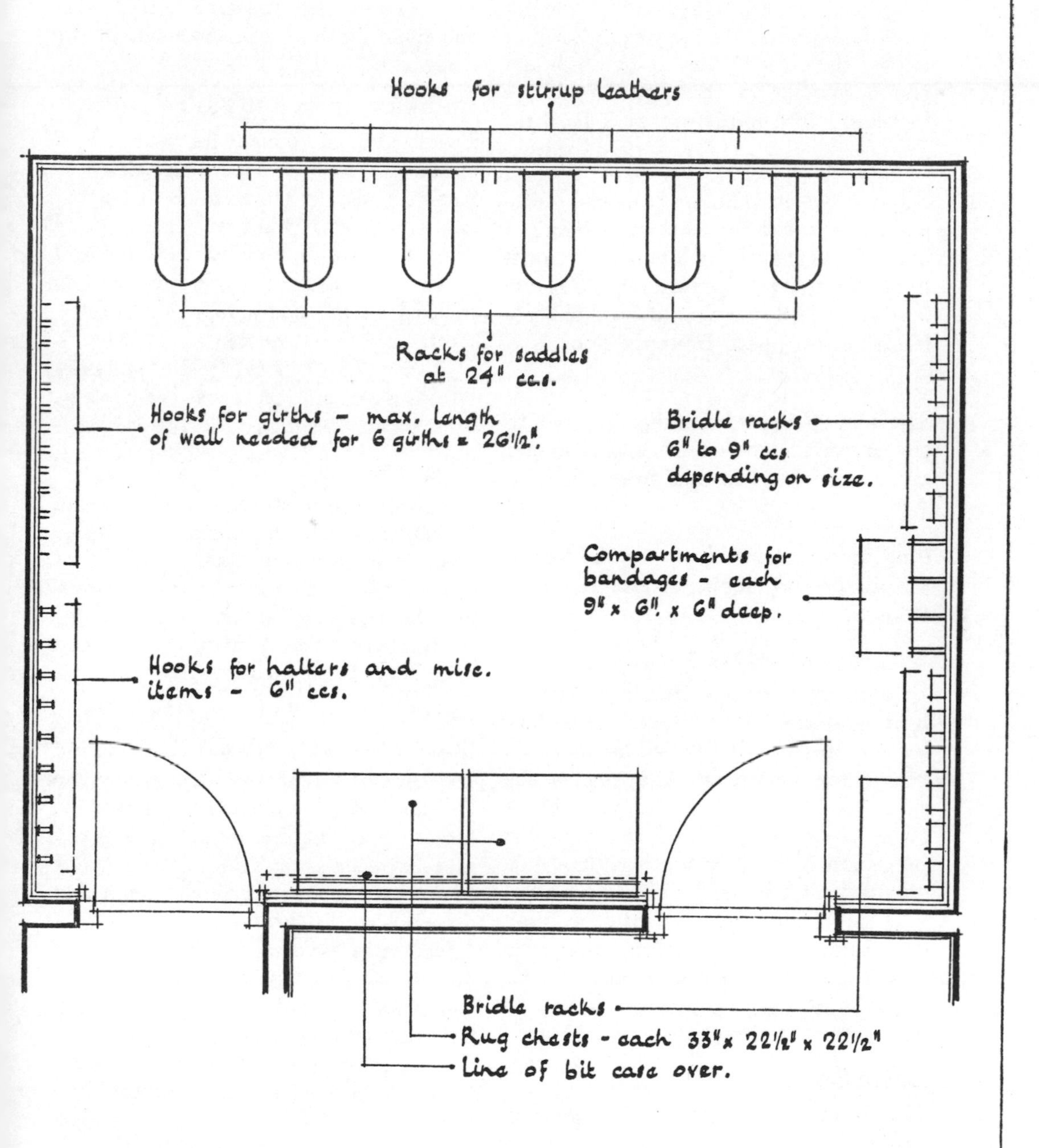

TACK ROOM — plan

rack is quite high enough for a person to reach comfortably. If the racks incorporate the bridle rack they must be fixed in one line only or in two lines staggered, and then adequate space must be left between the lower saddles to allow the bridles to hang between them. Spacing in this case should be increased to a minimum of 2 ft. 6 in.

Stirrups and Leathers

Stirrups are often hung at the end of the saddle rack with the leathers hanging to one side as shown, each pair on its own hook. To prevent damage or soiling the leathers the hooks should be fixed not less than 5 ft. 6 in. from the floor.

Girths

These are hung in a similar manner to the stirrup leathers and hooks should not be less than 5 ft. 6 in. from the floor.

Martingales

Similar conditions apply as mentioned for girths but hooks should be 6 ft. from the floor.

Bridles and Head Collars

Bridles are hung on standard pattern bridle racks which in some cases may be combined with the saddle racks as previously mentioned. Allowing for the reins to be caught up in the throat latch or on the hook of the rack the racks should be fixed not less than 5 ft. 6 in. from the floor.

Head collars (halters) may be hung on similar racks or on halter hooks. The height may convenientlʎ be at 5 ft. 6 in. and the lead rope if a long one can be coiled over the top of the rack or hook.

Lungeing Reins

The head-piece should be hung on a bridle rack at the height mentioned above, the rein which is 25 ft. long may then be coiled up over the rack, or on the hook which is often an integral part of the rack.

The above list covers the tack which will be found in the average hunting and hacking stable. There are, of course, many other items and the provision for storage must be agreed with the client at the commencement of the design.

Blankets, Rugs and Sheets

Provision must be made for the seasonal storage of clean horse clothing. These items are usually stored in chests, either free standing or built in. These chests must be strongly constructed, dry and be fitted with close fitting tops or lids rebated at the edges to exclude dust, moths and other vermin.

Standard galvanised steel chests may be obtained, a typical size being 2 ft. 9 in. × 1 ft. 10½ in. × 1 ft. 10½ in.

Bandages

Will also require storage space and can be kept in cupboards fitted in a suitable position or in pigeon hole racks. Allow space for four leg and one tail bandage per horse, plus about 10%. A leg bandage measures 5 in. wide and when rolled equals about 3 in. diameter.

Bit Case

A case for bits is sometimes required and the one illustrated was designed to accommodate the client's bits plus 20% for future requirements. The only method of arriving at a suitable size is to discuss the needs with the client and design accordingly.

The case is usually constructed of hardwood, is about 3–4 in. deep and the front is fitted with glazed doors. The bits allowed for in the case illustrated were hung on stainless steel hooks, positioned to suit the individual bits.

Repair Table and Storage

In most establishments, major repairs to saddlery will be sent to the saddler but small maintenance repairs will be

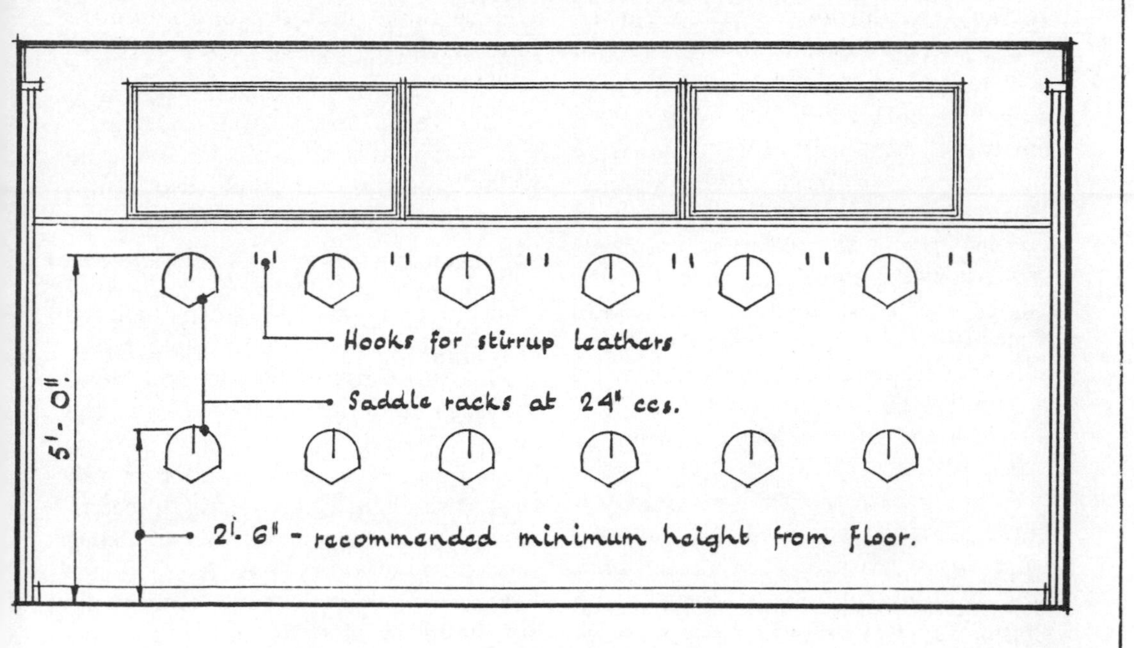

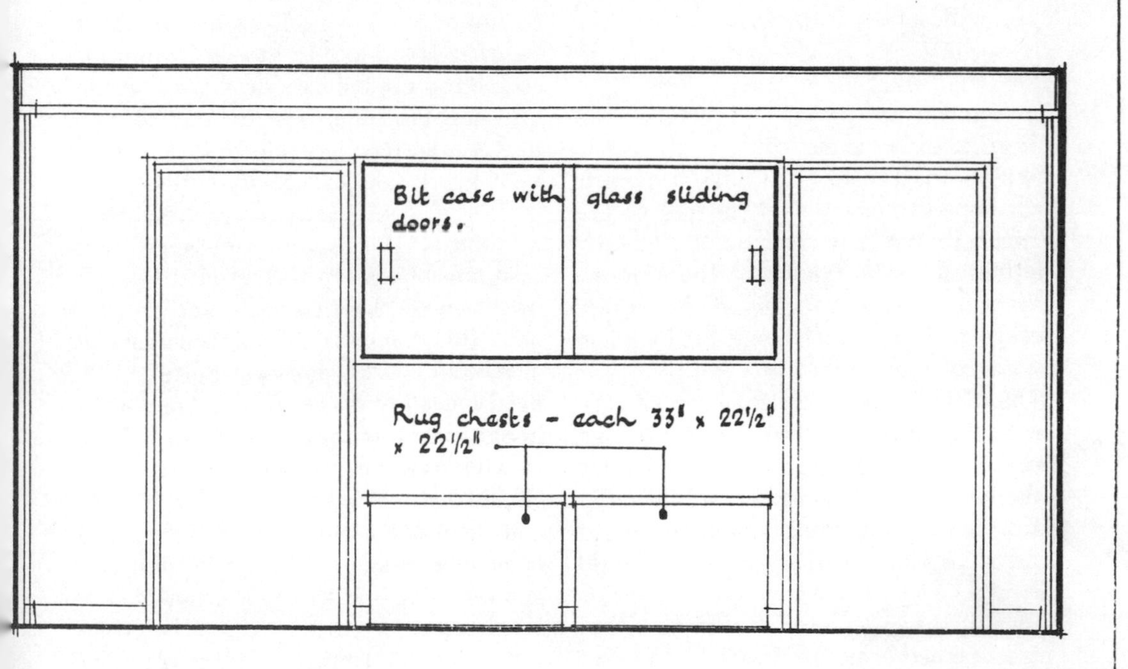

TACK ROOM — elevations

carried out on the premises, usually in the tack room. Provision should be made for this work.

A table, of stout construction about 4 ft. 6 in. long × 2 ft. 6 in. wide with a hardwood top, will serve adequately the purpose of most establishments. Drawers should be fitted and preferably divided to suit the tools and materials needed, twine, needles, punches, awls, knives, etc. The table must be placed in a position with good natural lighting and good artificial lighting will also be required, this will probably entail a separate and adjustable lighting unit in this position.

Some large establishments may require provision for more extensive repairs in which case a separate room may be required fitted out as a saddler's shop. The needs in such cases must be discussed with the client and the saddler. The necessity for a room of this type will, however, be rare.

Medicine and Poison Cupboard

Medicine and poison cupboards are often included in the furnishing of the tack room. The size of these cupboards will depend not only on the size of the establishment but also on the client's requirements in respect of the amount of equipment and medicines he wishes to stock. Some stock very little, some a considerable quantity.

Medicines and poisons must be separated into separate cupboards and each must be clearly and permanently labelled. The cupboards are usually hung to the wall at a convenient height and are about 6 in. deep. Shelves should be of glass or plastic covered. The inside faces of the cupboards should be either plastic faced or painted with gloss paint to facilitate cleaning. Glass or solid doors may be fitted as preferred. The doors to the poison cupboard must be fitted with a strong lock.

A table or bench with a plastic covered or glass top may be required to be fitted adjacent to the cupboards to enable medicines to be mixed. If this is required it is an added advantage to fit a small sink about 16 in. × 12 in. × 6 in. supplied with both hot and cold water. A power point should also be provided to enable a small sterilizer to be fitted.

Similar requirements in respect of artificial and natural lighting will be required as discussed for the repair table.

The above list of fittings will cover the needs of most tack rooms, even the larger ones. In addition some storage space in the form of either floor or wall cupboards should be provided to accommodate the many small miscellaneous articles requiring to be stored. The unused areas of wall space could with advantage be used for this extra cupboard space.

Great care must be taken over the design of the finishes for the room itself. Leather is subject to attack by mildew so the room must be warm, a minimum and constant temperature of about 55 degrees is recommended. It must be ventilated adequately and the possibility of condensation reduced to a minimum. At the same time consideration must be given to the hard wearing qualities of the materials and finishes used and the ease of maintenance and cleanliness. The racks require to be firmly fixed and as additional racks and hooks may be required at future times, materials which are difficult to fix to should be avoided.

To comply with the need for warmth it is recommended that the structure should comply with the thermal insulation values recommended for the boxes. Ceilings should preferably be of insulation board which will not only assist in obtaining the required U value but will also assist in reducing condensation. Walls may be plastered and painted, though hard setting coats should not

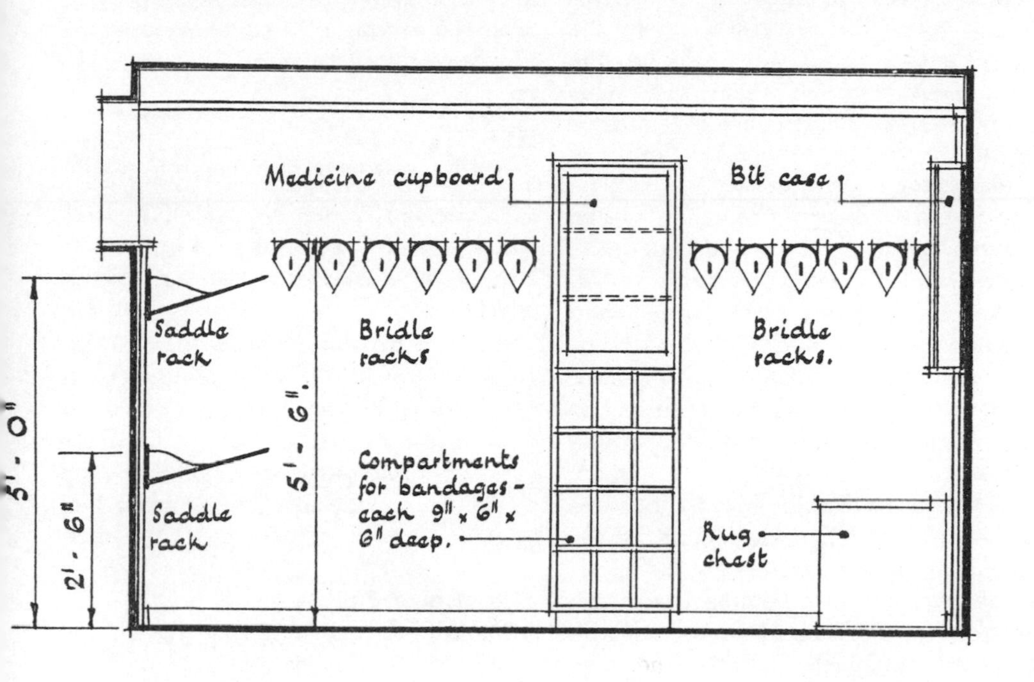

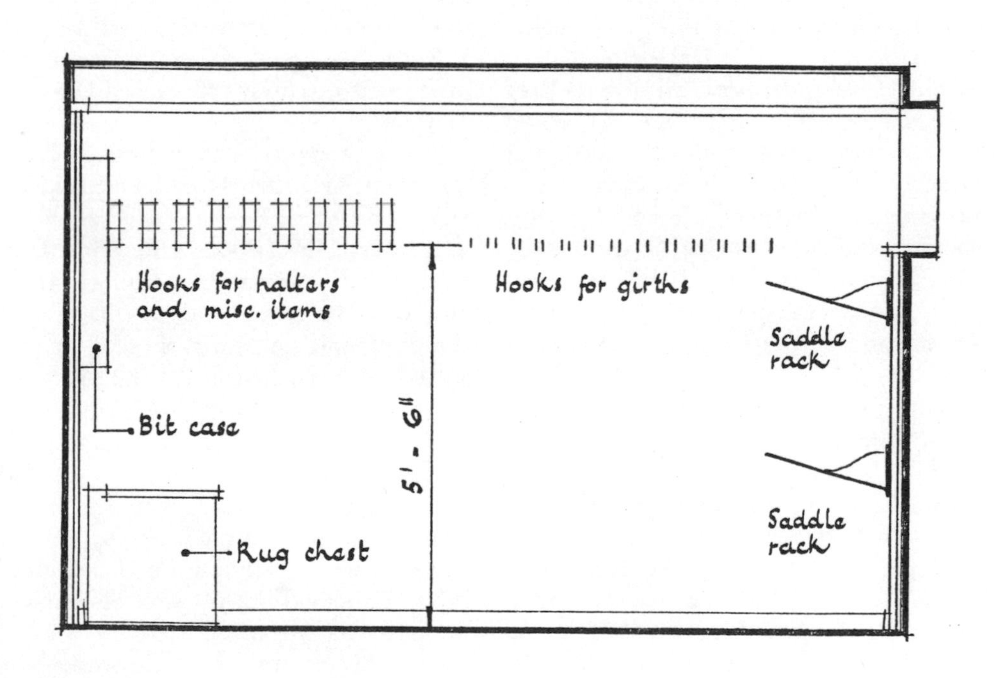

TACK ROOM — elevations

be used as this will encourage condensation to form. The old method used for finishing tack room walls was to line them with match boarding, this forms an excellent surface and is simple to fix to. A timber lining is recommended and many types may be obtained nowadays, all of which will be suitable providing there is an adequate thickness of timber for fixing purposes. This is the one room in the stable group in which practically any type of flooring, providing it stands up to hard wear, may be laid, so the architect is left in this instance to his own discretion.

As this room should be warmed, some suitable form of heating should be installed as a permanent provision. If central heating is available and pipes and radiators can be fitted, ensure that they are fitted well clear of saddlery to prevent both damage and undue drying of the leather. Similar requirement will apply to electric radiators and other forms of heaters fixed at or near floor level. If electric fires are used they should be fixed at high level and placed so that direct heat from them does not fall on leather work. Probably the most satisfactory method of heating this room would be by means of fan heaters. This type of heater obviates any direct heat playing on the saddlery and only warms and circulates the air within the room.

This room will require both lighting points for the repair table and bench if provided, and also a point or points for normal lighting of the room. Fluorescent strip lights are recommended as they give an overall evenly distributed diffused light. The one power point mentioned adjacent to the bench will serve the needs of this room adequately. Although there is little danger of water being used, a continuation of the waterproof installation is recommended. Any switches or points adjacent to the bench, if a sink is fitted in connection with the mixing of medicines, must be of a waterproof type.

WASHING AND CLEANING ROOM

(See illustration facing page 58)

The cleaning room illustrated is that which has been incorporated in Layout No. 1 and contains the following equipment:

(*a*) A sink 36 in. × 18 in. × 10 in. with a hardwood draining board which has been extended to form a bench, which may be used for cleaning purposes. Cupboards, for the storage of cleaning materials have been fitted under the bench top.

(*b*) One saddle horse.

(*c*) One drying rack.

(*d*) One bridle cleaning holder.

This amount of equipment will be sufficient in this room for most small establishments, larger ones will require at least the same basic equipment but in greater quantity, or size as the case may be.

The sink shown is recommended as a minimum size and should be provided with hot and cold water. The bench top must be of hardwood, preferably teak as it requires to stand up to both water and hard wear. The wall at the back of the sink and benching should be protected with a skirting of tiling at least 12 in. high. The under-bench cupboards shown are standard EJMA floor units and are quite suitable for the storage of cleaning materials.

A boiler if required may be of any suitable size depending on the amount of use it is likely to receive. Sheets and bandages, cloths, etc., are invariably washed on the premises but blankets are more often dry cleaned nowadays than washed.

The saddle horse shown is a standard horse 4 ft. 6 in. long and is moveable.

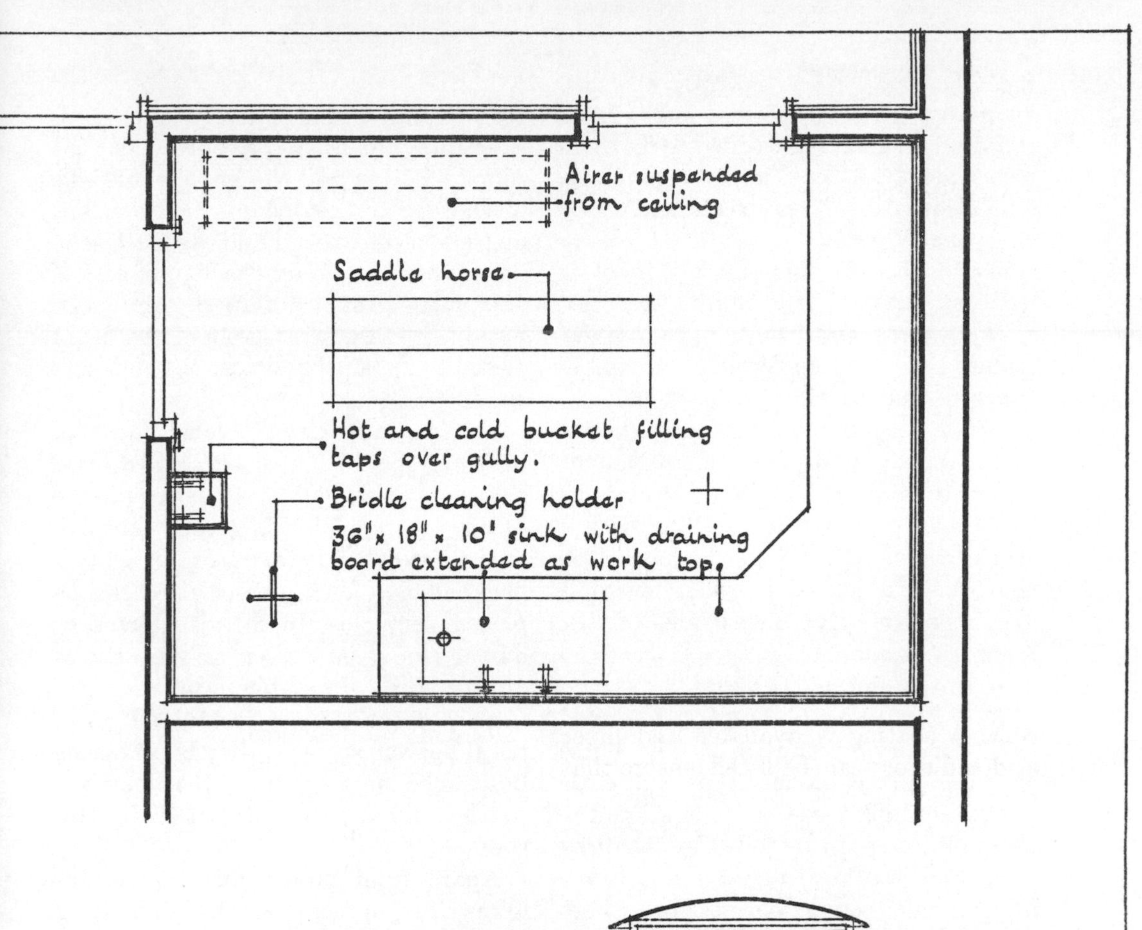

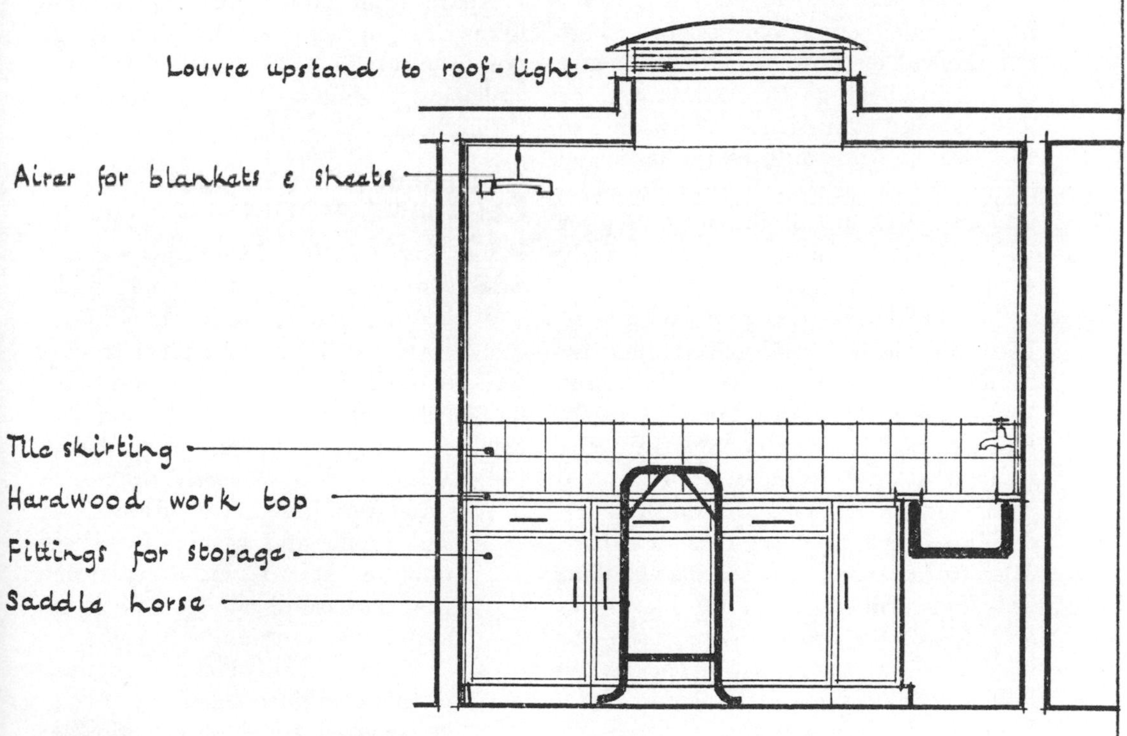

CLEANING ROOM

In many cases more than one will be required and if it is intended to fix them permanently in position at least 2 ft. 6 in. clear space should be allowed round each one. Where they are fitted side by side allow a minimum of 4 ft. 6 in. clear between them. The horses are not usually fixed and where they are installed as moveable furniture adequate space should be allowed for them.

Bridle cleaning holders may be obtained as standard pieces of equipment, either to a fixed length or telescopic. They are secured to the ceiling with a strong hook and should be placed so that there is adequate room for their use, if possible give a clear area of 2 ft. 6 in. all round. They should be fixed near to the sink if possible.

The clothes airer shown is intended both for drying and airing the horse clothing. Fix it so that it is well clear of the cleaning areas and not where the clean clothing can be soiled by the dirty tack, etc., waiting to be cleaned. Don't hang it over saddle horses, etc., where the washed clothing can drip onto any of the saddlery as the leather will be marked and stained. An alternative to the airer shown would be the provision of an electric airer, and it would be an advantage to install this type of airer where heating cannot be adequately supplied in any other way.

The finishes to this room require to be easily cleaned, and glazed tiling for lining the walls is probably the most satisfactory. The floor must be easily cleaned and also be capable of being washed down. Tesselated or quarry tiling are very suitable finishes. To facilitate washing down it is an advantage to fit a gully, and slope the floor finish towards it. The gully shown on the illustration serves the provision for washing down, the emptying cock to the boiler and the bucket filling tap.

Care must be taken to provide adequate and preferably cross ventilation to this room. There will be considerable humidity present when the room is in use and the vapour must be discharged to the open air before it has an opportunity to build up and seep into the adjoining tack room. If adequate cross ventilation cannot be provided an extract fan should be fitted close to the boiler if fitted, the sink and the drying racks.

To facilitate drying some form of heating should be fitted. If central heating is available, coils could be fitted at high level. Other suitable heaters would be radiant or fan heaters fitted at high level. These should not be placed very close to the airing rack or damage may be caused to the articles on the airer. Probably forced air fan heaters are the most suitable. If electrical heaters are used they must not be positioned near to or on the floor due to the amount of water used in this room.

Apart from points for the various items of equipment mentioned, lighting points only will be needed and the installation should be waterproof throughout.

COMBINED TACK AND CLEANING ROOM

It has been mentioned that many small establishments will require their tack and cleaning rooms to be combined. Some form of compromise must therefore be worked out in such cases. The details of both rooms have been fully discussed and the following points have been enumerated to assist the architect in reaching a satisfactory solution.

1. The drying and washing facilities must be kept as far away from the clean saddlery as possible, to avoid splashing and soiling it.
2. A boiler should not be installed in this combined room and if one is required some other suitable position should be found for it.

3. Drying racks should preferably be outside the room but where fitted within the room they should be fitted close to a window at high level to avoid increasing the humidity more than necessary.

4. The floor should conform to the requirements of the washing and cleaning room and not be of the types suggested for a separate tack room.

LITTER DRYING SHELTER

The position of this shelter should, as mentioned in Part 1, be close to the stable yard. In large establishments more than one shelter may be required.

The size of the shelter will depend on the number of boxes it serves and the use made of those boxes. Before any decision is reached on the design requirements the matter should be discussed with each individual client, for as with the other details given in this book, the bedding of horses will vary from one stable to another.

As previously mentioned the litter should be spread on a grillage raised above floor level to allow a free circulation of air around it to accelerate the drying time. This grillage should be of stout construction and be of a non-absorbant material. Steel or concrete grilles are recommended, particularly the former which being lighter may be arranged and fitted to be easily removable and therefore facilitate cleaning. Such grilles may be set in rebates on the upstands or be hinged. It is recommended that such grilles be fitted at least 12 in. above floor level but where fixed grilles are provided a greater height above the floor will facilitate cleaning underneath them.

The main floor to the shelter should be of concrete or other smooth impervious material for ease of cleansing. It should be laid to falls to allow for drainage when washing down, and the gully taking the effluent should be a mud or similar type gully to prevent any straw entering the main drainage system (see chapter on drainage).

It has been mentioned in Part 1 that the roof of this shelter should be formed to overhang sufficiently to prevent driving rain wetting the litter. This is the best method as then there is no obstruction to the free circulation of air. However, in some cases where space is limited the additional area required for such an overhang will not be available. In these circumstances it will be necessary to protect the litter by means of louvred screens. These screens should be of stout construction but must at the same time be easily movable so that they may be positioned on the windward side of the shelter. For small establishments with only a small shelter the screens may be formed with wheels set to allow an overlap between the panels. An alternative but more expensive method is to form the panels as sliding folding screens set on a track, in a similar manner to sliding doors to a garage. In such a case the track should extend as a continuous railway around the shelter but the panels need only extend for about half the overall circumference. If this method is used the panels should be hung from an overhead track and be guided at the base by a Tee section guide as recommended for other sliding doors throughout the stable unit.

It is unlikely that there will be much demand for a shelter of this type at the present day. Few, if any, small establishments will require one and the large establishments will be more likely to install a power dryer. The latter will be a specialist provision and the design must be worked out in consultation with the manufacturers of the power unit.

MANURE STORAGE

In most cases the areas for the storage of manure are formed as open bunkers with brick or non-absorbent block walls on four sides. An opening about 3 ft. wide is normally left in one side to facilitate emptying. The floor should in all cases be formed of concrete to prevent the ground becoming foul. To prevent the surrounding ground being fouled by the water discharging from the bunkers some form of drainage must be provided, this may take the form of a gully set in the base of the bunkers, though this is very liable to become choked. A second and more suitable method of drainage is to slope the floor of each bunker towards the opening and set the gully outside the opening.

The size of the bunkers provided will depend both on the number of horses and the arrangements made for the collection of the manure. It is is difficult to give a rule of thumb for the size of the bunkers as the compaction rate of manure is rapid and will depend on weather conditions. During a number of experiments it was found that a 6 ft. high stack, during very wet weather reduced to between ½ and 2/3rds of its original height, within one month.

It is necessary however to give some guide to the architect on the subject so that he may be able to design as accurately as possible. The following conclusions have been reached after somewhat lengthy experiments had been carried out.

Storage for one week allow 7 cubic yds. per horse.

Storage for two weeks allow 12 cubic yds. per horse.

Storage for three weeks allow 16 cubic yds. per horse.

Storage for four weeks allow 20 cubic yds. per horse.

The above schedule allows for a slightly generous storage volume as no compaction has been allowed for during one seven-day period. The schedule has been based on combine straw only as this is the most common type in use at the present day.

MISCELLANEOUS STORAGE

In addition to the storage units already discussed, storage space must be found for the various pieces of equipment, i.e., wheelbarrows, forks, shovels, etc., which are used around both the boxes and other parts of the group. One or more stores will be required dependent on the size and the layout of the establishment. Generally a covered recess of suitable size as shown in Layout 1 will adequately serve the purpose for the smaller establishment. A large establishment may require a fully enclosed store or stores but often the open type shelter will satisfactorily serve its needs. The storage area must be of sufficient size to accommodate all the barrows, and racks should be fitted to the walls so that the utensils may be hung up.

Where ranges of boxes are provided with an overhang to the roof to form a covered way it may not be found necessary to provide storage. Racks may be fixed at convenient points on the walls and the utensils hung on them will be protected sufficiently from the weather. In cases where "Utility" boxes are provided, as shown in Layout 2, each groom will use his box for the storage of his tools and equipment.

The size of the store for these articles will best be decided in consultation with the client. As a rough guide allow storage for one barrow, one pitch fork or hay fork, one shovel and one broom for each groom.

The position of the store should be off the stable yard in all cases as it is in this area that these pieces of equipment will be mostly used.

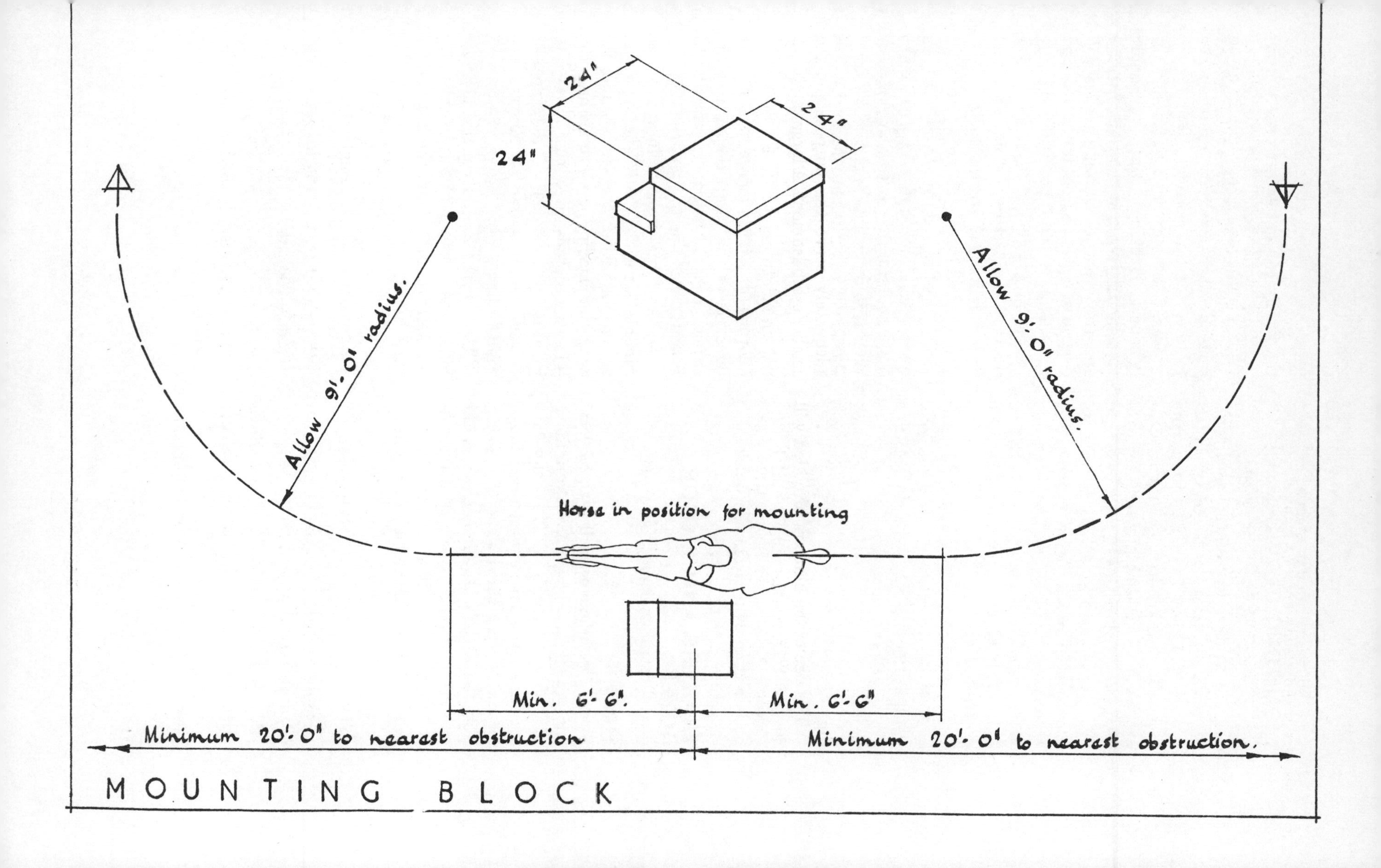
24"
24"
24"
Allow 9'-0' radius.
Allow 9'-0" radius.
Horse in position for mounting
Min. 6'-6".
Min. 6'-6"
Minimum 20'-0" to nearest obstruction
Minimum 20'-0" to nearest obstruction.
MOUNTING BLOCK

MOUNTING BLOCK

(See illustration facing page 62)

The illustration shows a suitable form of mounting block consisting of a platform formed of a 24 in. × 24 in. paving slab supported on brick walls at a height of 24 in. with one step for convenience of mounting. The step should always be positioned as shown in relationship to the horse when in the mounting position. A free standing block capable of being mounted from either side should have two steps, one on each side.

The position of the block should allow a horse to be brought up to it in a straight line and allow the rider to ride off in a straight line. The full length of the horse should be clear of the block before a rider needs to turn him and such a turn should be on a gentle curve as illustrated. The nearest obstruction to allow of these conditions should not be less than 20 ft. from the block.

TRAILERS AND MOTOR BOXES

(See illustration facing page 64)

Trailers

Most establishments will own at least one trailer and accommodation for it will often be required. This may take the form of an entirely enclosed building or an open sided structure. Ascertain the client's requirements and the number and types of the trailers to be accommodated. The trailer in the smaller establishment may be towed by the family car but larger ones will invariably own a Land Rover or similar type of vehicle for trailing purposes. In such cases accommodation may also be required for this vehicle.

Adequate space must be allowed for turning, backing and generally manoeuvring the combination of trailer and trail vehicle. Turning circles will depend to a great extent on the size and type of trailer used, but in the main by the towing vehicle. The allowances required must therefore be calculated to suit individual circumstances.

Proper allowance must be made for loading and unloading the horses both by rear and by side ramps. A horse must be brought up to the ramp, or leave it in a straight line, so the animal should not be turned at a lesser distance than one length clear of the end of the ramp. A 16 hhs horse measures about 8 ft. in length overall from nose to tail, allowing for a fairly gentle turn into the direction of the ramp the nearest obstruction should be not less than 22 ft. from the ramp end.

To assist the designer in arriving at suitable sizes for both the building, to accommodate the trailer and for areas for loading and unloading the following details give the up-to-date information on some of the types of trailers manufactured by Messrs. Rice Trailers Ltd.

These types are only a few of those made and accommodate, in order, for one and two horses and the last two for three horses.

Type	Width o/a	Height o/a	Length o/a
Beaufort Single	62 in.—157 cms.	104 in.—264 cms.	164 in.—417 cms.
*Beaufort Double	86 in.—218 cms.	105 in.—266 cms.	164 in.—417 cms.
Beaufort 3 Standard	86 in.—218 cms.	105 in.—266 cms.	187 in.—475 cms.
Beaufort 3 H D	89 in.—226 cms.	106 in.—269 cms.	187 in.—475 cms.

* Shown in the illustration facing page 64.

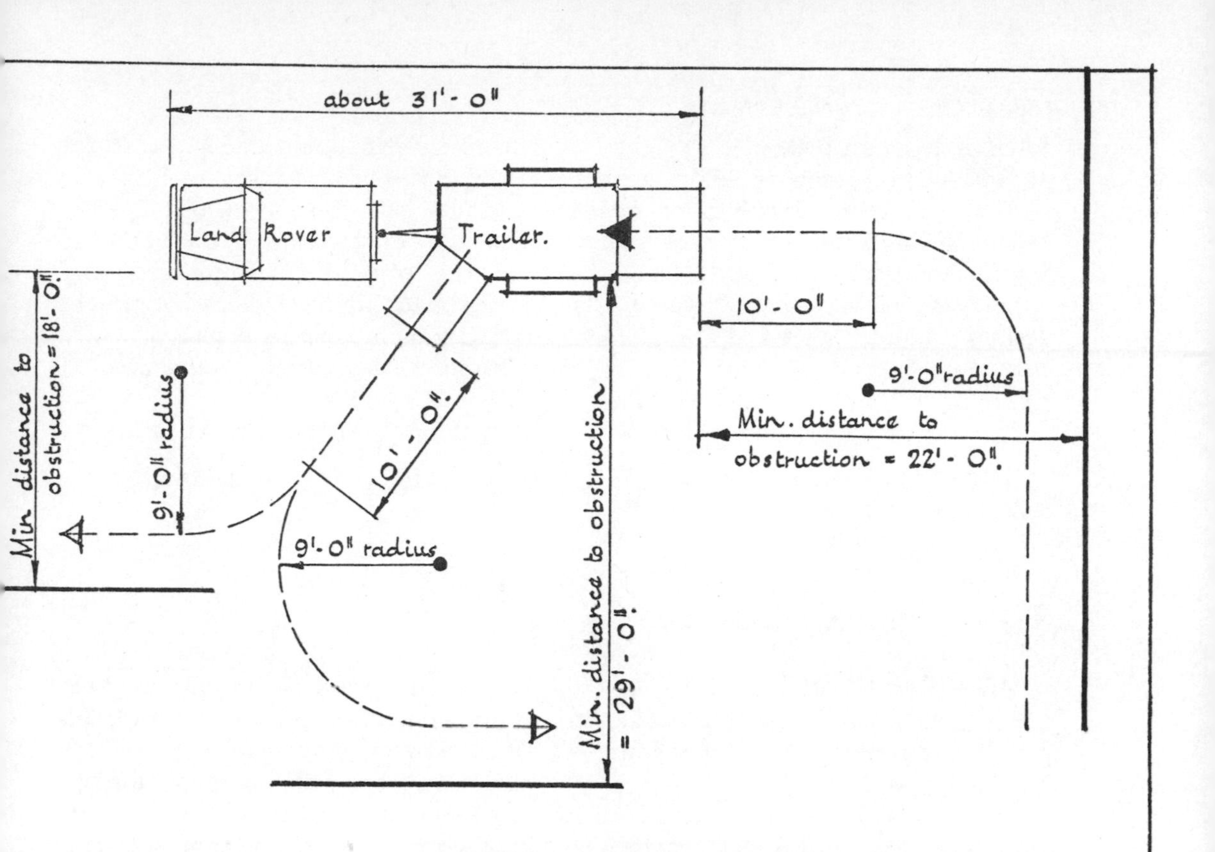

TRAILER The recommendations indicated are based on the Beaufort Double Trailer.

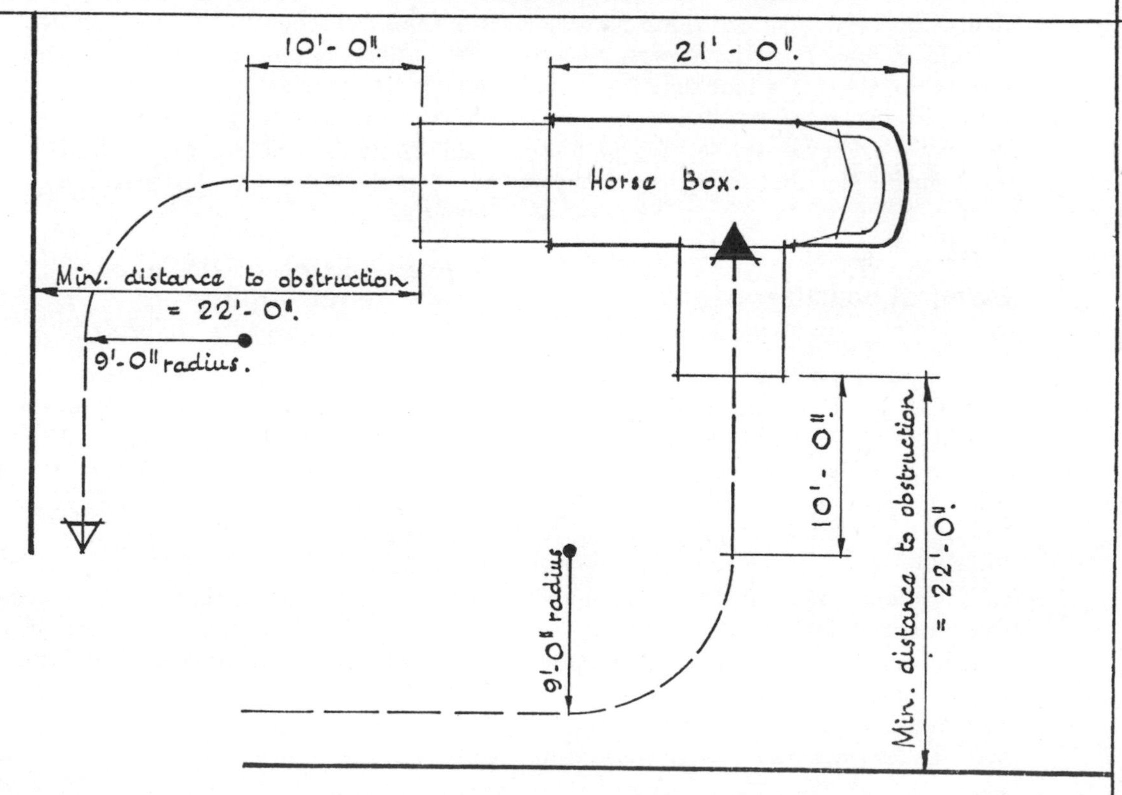

HORSE BOX The recommendations indicated are based on the Warwick (3 horse) Box.

Motor horse boxes

It will be necessary to make similar allowances, when planning for motor boxes. To assist the designer the following information, based on three of the types of boxes, manufactured by Messrs. Charles Clark Ltd., of Wolverhampton, is given.

	Width o/a	Length o/a	Turning Circle
Dudley—2 horse	7 ft. 5 in.	18 ft. 10 in.	41 ft. 6 in.
*Warwick—3 horse	7 ft. 5 in.	21 ft. 0 in.	47 ft. 0 in.
Kenilworth—4/6 horse	7 ft. 5 in.	27 ft. 0 in.	60 ft. 0 in.

* Shown in the illustration facing page 64.

Allow about 7 ft. 6 in.–8 ft. for the projection of the ramps, both side and rear, when they are in the lowered position.

In cases where a client has his own box this should be measured, or when he intends to buy a new one find out the type and make and obtain the up-to-date information from the manufacturers. Remember that even when clients do not own a box that these are often hired for a day, so some provision should be considered for loading, unloading and manoeuvring the vehicle.

FIRE PRECAUTIONS

General considerations

It will be appreciated even by a person not conversant with stable management that the danger of fire in stable buildings is a very serious problem. Much can be done during the design stages to reduce the danger, particularly of a fire spreading. This problem should be in the forefront of an architect's mind when considering both the planning and the constructional details of the scheme. Considering the highly inflammable nature of much of the material stored and used, the lack of thought given to this matter in many establishments is surprising. No doubt much cause for laxity in this respect is due to economic considerations, but even if only the cash value of the animals, the buildings and the contents is considered, any saving made at the expense of proper fire protection is to be deplored.

Riding schools, letting out horses on hire are covered by the conditions of the Riding Establishments Act 1964. This Act, however, only requires that "appropriate steps will be taken for the protection of horses in case of fire" and goes no further. No recommendations are made as to how this protection is to be given, in fact the Act only nibbles, and that but slightly, at the outside edge of the problem and makes no attempt to tackle it with decision. There are in fact no Acts which cover stable buildings and each is dealt with as a special case by the Authorities concerned.

The prevention of fire and escape requirements

Most architects will be familiar with the basic principles of fire precautions. There are however a number of factors relative to stable buildings which may not be immediately obvious to the designer. It is therefore felt that it is necessary to devise suitable recommendations which if followed will materially reduce the fire risk. These recommendations should be read in conjunction with the Means of Escape Regulations of the appropriate Authority in whose area it is intended to build.

Hay and straw storage

The main fire hazard within the total stable group is in the hay and straw storage areas. Fires in these stores can be caused by overheating of the bales or by accidental ignition, i.e., smoking or faulty electrical apparatus. The first mentioned cause can only be overcome by the owner but can be assisted by the architect in providing a properly ventilated building giving a continuous flow of air throughout as already described. Smoking of course should never be permitted in or near these stores but this cause which is probably the most common cause of fires is outside the control of the architect. Faulty apparatus can be overcome by the provision of a properly designed installation, regularly tested and maintained. As fires have been known to be caused by hay and straw being stacked close to electric light fittings, care must be taken to ensure that such fittings are well clear of stacking areas.

Whilst discussing the planning of the buildings it was stated that from the point of view of convenience and economy of labour these stores should be situated close to the stable buildings. These stores should, to reduce the fire risk be positioned well away from other buildings. Obviously therefore a compromise will often need to be reached so that all factors are adequately catered for.

A minimum of one hour's fire resistance between the buildings is recommended and this degree of resistance may be provided in basically two ways. One, by a structural division of one hour's fire resistance or by an adequate fire break giving not less than the above requirement. If a fire break is formed, a minimum width of 15 ft. is recommended and this area must be kept free at all times of inflammable materials. It is assumed for the above recommendation that buildings will not exceed 15 ft. in height to the eaves and be of single storey construction. No doubt many architects will point out that this width exceeds the normal recommendations, this has not been overlooked but the recommendation still stands.

If a structural division is formed it must be imperforate and care must be taken to ensure that a fire has no opportunity to "jump across" this division by means of either the roof or through adjacent openings. It is recommended that no openings be formed within at least 3 ft. of each side of the main dividing wall. Where roofs abut the division they should be of one hour's fire resistance for at least 3 ft. on either side.

Where hay and straw stores are situated in a loft over the loose boxes of other buildings similar recommendations apply in respect of the structural division (in these cases the floors) and of openings. Owing to the design load required to accommodate the material it will in most cases be found to be of adequate strength to support the loft over if it should collapse. This possibility must however be taken into account at the design stage and allowance should be made for such an eventuality. In addition to these provisions it will be necessary to insulate the floor to prevent the heat seriously affecting the horses prior to their evacuation. Such insulation is probably best formed by a suspended ceiling of Celasbestos, suspended on metal hangers beneath the floor. Ensure that the hangers and any other supports to the ceiling are of fire resisting material. Any windows formed in the outer walls of the loft over the doors of the boxes or the escape passages should be fixed lights of at least 1 hour's fire resistance. This will prevent burning material falling either into the boxes if the upper parts of the doors are open, and thus spreading the fire, or into the passages and obstructing the escape routes.

Where the ventilation system incorporates the use of ducts care must be taken to ensure that the construction of the ducts, their positions and the positions of inlets and outlets, do not assist the spread of the fire. Ducting should if possible be run clear of all areas of high fire risk, but where passing through such areas protection must be given of at least 1 hour's fire resistance.

Feed storage

The feed store can also be an area of high fire risk, particularly where the corn is stored in bulk and internal combustion can take place. Similar precautions are therefore recommended as discussed above for the hay and straw stores.

Loose boxes

Fires starting in loose boxes will in most cases be caused by the carelessness of a person smoking. The recommendations so far discussed have been based on the primary factor of keeping a fire away from the horses and thus allowing their evacuation or even leaving them safely in their boxes while the fire is dealt with. It is of even greater urgency to put out a fire starting in a box quickly and to prevent it spreading. Most fires starting in the boxes will take place during normal working hours and can be put out quickly as the staff will be on hand. The only adequate provision that can be made to deal with such a fire quickly when the staff are not available is to install a suitable sprinkler system actuated by a fusible link control. It is normal for such systems to be installed with the sprays set below the ceiling. An alternative, relative to this problem, would be to install the sprinklers at low level around the walls of the boxes so that the litter only would be sprayed. In either case the control would need to be situated at low level in each box. The sprinklers must be protected from damage by the horse, either, if at ceiling level, at a height of not less than 8 ft., or if at low level they should be recessed in the box walls.

The other units in the group

The other units in the stable group are not generally of high fire risk providing adequate control is exercised over the appliances used, particularly oil heaters.

Minimising risk by sub-division of boxes

In large establishments it is recommended, that to minimise the fire risk the ranges of boxes be divided into groups by means of either fire breaks or divisions. Groups should not exceed twelve boxes. Such divisions are particularly necessary where a large number of animals are housed under one roof as shown in Layout 3. This matter is dealt with to greater length later in this chapter, when the precautions are discussed relative to Layouts 1 and 3.

Means of escape

The means of escape requirements will depend on the size of the establishment, the siting of the group and the planning of the units within the group and their relationship to each other. In the case of a range of loose boxes each box opening into a yard or wide passage, the requirements are fairly simple. The yard or passage must be kept free of inflammable materials at all times and in the case of a passage it must be of suitable width. A minimum of 10 ft. in clear width is recommended, unless the passage also acts as a fire break, when 15 ft. clear is recommended. The provision of these recommendations will enable the horses to be brought out quickly and taken to safety with the minimum of trouble. Alternative routes for escape purposes must be provided to avoid the necessity of taking the horses past the area of the fire. In the country or where the stables adjoin a paddock it should be possible to take the horses direct to the paddock. Where no paddock is available, as is often the case with stables situated in a town, temporary accommodation will be needed elsewhere. Each owner should consider the possibility of a serious fire and decide on what action will be needed should it become necessary to evacuate the horses. The staff should be informed of any arrangements made and printed instructions covering such an eventuality should be displayed. If adequate forethought is given to this matter there will be less confusion if a fire should occur.

Where the stable buildings are planned in the manner shown in Layout No. 3 the means of escape requirement may become more complex. There must be at least two exits to the building and each must be of adequate width (minimum clear opening of 6 ft.) and they must be kept clear at all times not only of inflammable materials but of any other obstruction. It is considered that the provision of two exits is only sufficient for accommodation of up to 12 horses. Over that number more exits should be provided and the following relationship of escape exits to the number of horses accommodated is recommended:

Up to 12 horses	2 exits minimum.
12 to 24 ,,	3 ,,
24 to 36 ,,	4 ,,
36 to 50 ,,	5–6.

It should be appreciated that horses are considerably frightened by fire and it is often difficult to evacuate them. To be strictly correct and in accordance with recommended practice each should be led out separately, so the greater the number of exits available, the quicker and easier will be the task of the staff. It is sometimes found in practice that a horse will refuse to leave the box and in such cases may have to be driven out. The widths of the exits and the swings of the doors discussed later have taken this possibility into account.

Escape doors

Doors to the escape passages must open outwards and be fitted with fastenings easily actuated by hand from either side. In cases where the boxes have been sub-divided into small groups, the doors to such groups in addition must be of 1 hour's fire resistance and be made self closing. When opening into a passage they should be hung so that they do not obstruct the width of the passage. Doors to the boxes should open into the line of the main escape route and should fold back so that they do not cause an obstruction and materially reduce the recommended clear width of the passage.

Fire check doors

When buildings adjoin, the doors between those of high fire risk and areas leading to the boxes should in addition to the other precautions discussed be of 1 hour's fire resistance and be rendered self closing. Doors between areas of high fire risk, i.e., the hay and straw stores, should also conform to the above recommendations. If it can be afforded it is better to fit such doors with a fusible link control.

The relationship of buildings to facilitate fire fighting

If a fire starts it will be necessary to ensure that not only the establishment's own fire fighting equipment can be brought effectively into use, but also that the brigade's equipment can be brought close to the fire. This factor must be considered during the planning stages and not be left for future consideration. Access roads to the buildings must have a minimum width of 10 ft. and be capable of supporting a weight of 10 tons. A road of this specification must be provided to within 120 ft. of the buildings. Most stables, particularly large establishments will require an access road up to them of adequate strength to support a horse box and the additional requirements of the road to support fire fighting vehicles will not add greatly to the cost of the road if allowed for at the time of laying it.

Fire points

However small the establishment may be at least one fire point should be provided. A recommended type of fire point is shown in the illustration facing page 70, and consists of a tank of 100 gallons capacity, fitted with a cover and hinged lid. In addition two 2 gallon buckets, one hand pump, and one metal sounder and striker should be provided. Replenishment of the tank must be provided for, so a hose of adequate length to reach the nearest stand pipe must be fitted. The layout shown incorporates a notice board on which the detailed instructions to the staff of the action to be taken in the case of a fire should be displayed. The buckets provided for use at the fire point should be of the round bottom type fire bucket, to prevent them being used for other purposes and consequently finding that they are not available in case of a fire. The tank must be protected against frost.

Fire points of this type will be adequate for small establishments, large establishments however will require greater protection. Where the water is of sufficient pressure, a fire hydrant conforming to BS 750, Standard Form 3/1/1/1 should be provided within 300 ft. of the buildings. In addition hose reels will be required to conform to BS 750 Standard Form 3/2/3/1.

Fire points and hydrants must not only be positioned in the manner discussed above but must be well clear of all buildings, particularly those of high fire risk, so that they will be available for use during a fire without unnecessary risk or discomfort to the staff. Suitable positions relative to the buildings are shown on the illustrations facing pages 73 and 75 which will be discussed later.

Fire alarms

The fire point recommended for small establishments includes a hand operated fire alarm. Large establishments are advised to install an electric alarm system in addition, with suitably positioned control points throughout the entire range of the buildings. An automatic alarm is recommended for night operation. Alarm bells to the system should be positioned at points within the curtilage of the layout and for night use in at least two adjoining houses if possible. Ensure that a telephone is available in each of the selected houses, so that early warning may be given to the brigade. Electric alarm systems must not be operated off the main electricity supply but should be separately powered by batteries, which should be properly maintained and frequently tested.

The main requirements of fire prevention and control relative to stables has now been covered and if read in conjunction with the Means of Escape and the Fire Regulations of the appropriate authority should ensure that the risk of fire is kept to a minimum and if a fire should occur that it can be speedily dealt with. The architect is advised to consult with both the County Fire Brigade Headquarters and with the client's insurance company as soon as the initial sketches have been prepared.

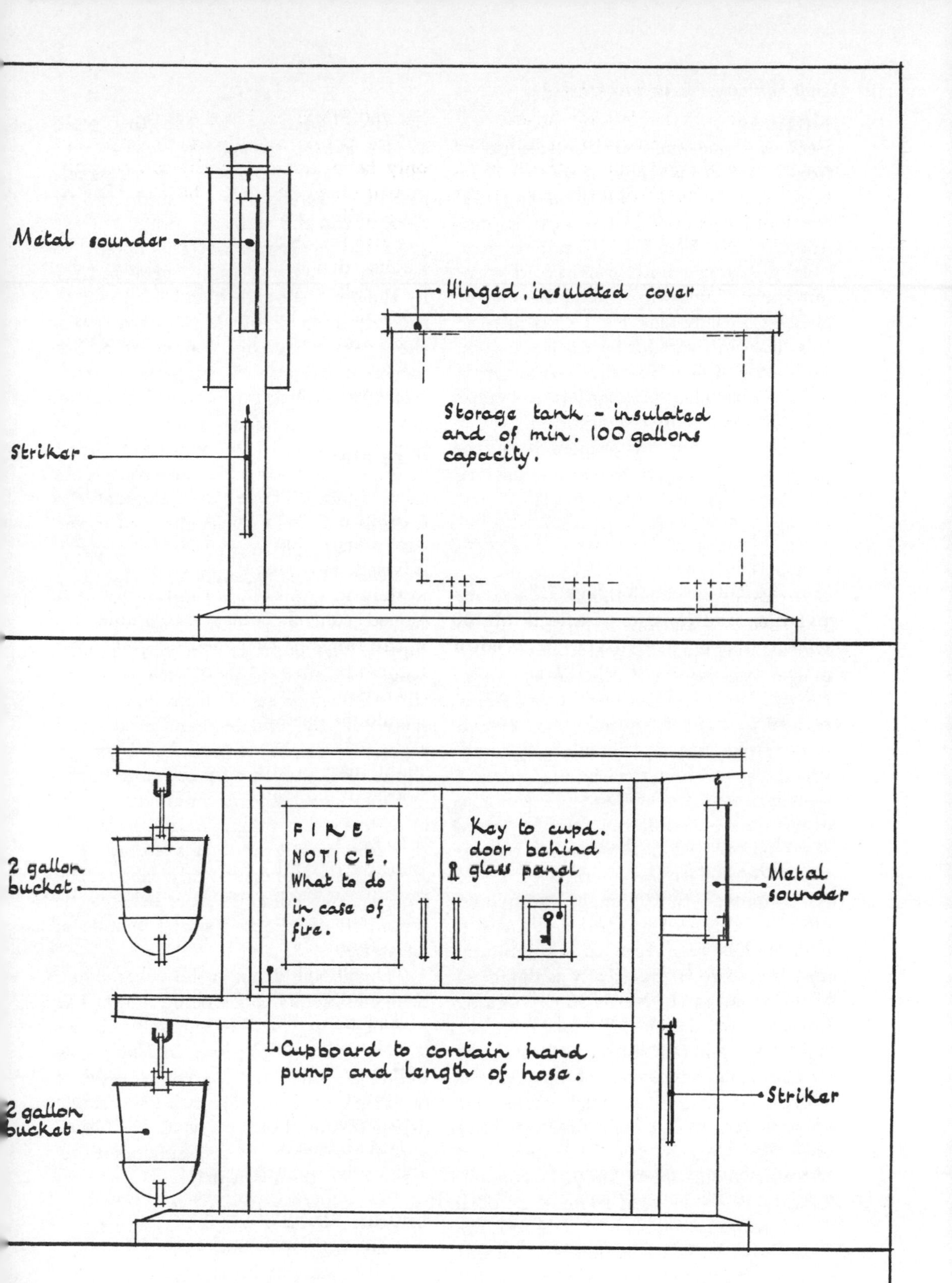
Metal sounder
Hinged, insulated cover
Storage tank - insulated and of min. 100 gallons capacity.
Striker
2 gallon bucket
FIRE NOTICE. What to do in case of fire.
Key to cupd. door behind glass panel
Metal sounder
2 gallon bucket
Cupboard to contain hand pump and length of hose.
Striker

FIRE POINT

Notices

Probably the most common cause of fires in stables is smoking, both by the staff and by visitors. It is essential therefore to have clearly written, permanent "NO SMOKING" notices suitably positioned both within and without the buildings. Such notices must be clear and easily read from a distance so the lettering should be at least 4 in. high. It will be found that red lettering on a white painted board is the most satisfactory colour combination.

Before leaving this subject it is proposed to discuss the recommendations in connection with Layouts Nos. 1 and 3.

Layout No. 1

(See illustration facing page 72)

The recommendations in respect of Fire Precautions are usually simple to apply in the type of small scheme shown in Layout No. 1 and this particular scheme proved to be no exception. The hay and straw store, which is the main fire risk, was positioned so that adequate cross ventilation was provided thus minimising as far as possible the risk of internal combustion within the stacks. In this project the client required the store to be connected to the main group and although this need was satisfied the connection has been kept to a minimum, and the store is practically a detached building, and well clear of the remaining buildings on the site. To minimise the risk of a fire in this store from spreading to the other buildings, the end wall as shown adjoining the feed room was constructed to 1 hour's fire resistance and the whole of the feed, tack and cleaning rooms were similarly treated. The doors to the feed room were both made self closing and of 1 hour's fire resistance. In this scheme the store was over 6 ft. higher than the remaining buildings so it was not considered necessary to change the construction of the roof adjacent to the fire wall. Ventilation to the side walls was provided by honeycomb brickwork panels and these were kept a minimum of 3 ft. clear of the end walls.

Should evacuation of the horses be necessary this was amply provided for by the gates shown which lead direct to the adjoining paddock. The two gates shown were fixed for convenience, but serve the purpose of escape very satisfactorily, as it allows two exits for the horses and therefore quicker evacuation. The doors to the boxes were hung in line with the main escape routes. As these doors all open back against the walls and there is plenty of room in the yard, this is not so important a point in this case as it is in Layout No. 3.

Two fire points have been provided in the scheme and were positioned as shown. The point in the yard is suitably positioned to deal with a fire in either boxes or adjoining rooms and is served by the two bucket filling stand pipes, one in the centre of each range. The second point is situated close to the hay and straw store but not sufficiently close to be uncomfortable to operate in the case of a fire in this store. It is served by a stand pipe adjacent to the garage, and could in this scheme be served by the house as well in case of necessity.

The existing drive in this scheme had been adequately constructed to take the load necessary for the brigade's vehicles to approach quite close to the stable buildings, and it was not considered necessary to make up the new branch drives to this standard, as all the buildings were within the recommended distance from the main drive.

Layout No. 3

(See illustration facing page 74)

This scheme raises a number of problems in respect of fire precautions which do not occur in the type of layout dealt

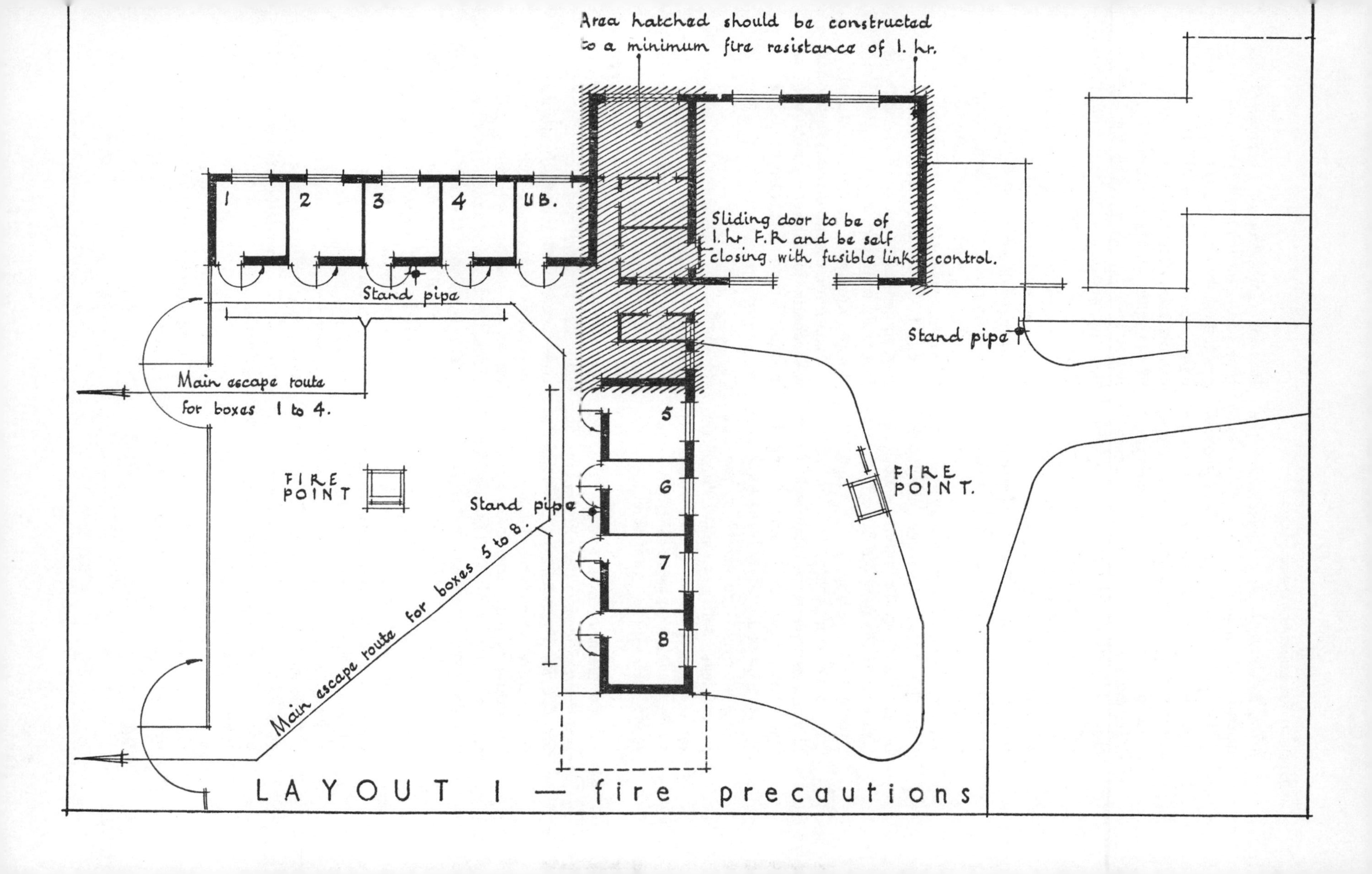

LAYOUT 1 — fire precautions

with in Layout No. 1, whatever the overall size of it.

The hay, feed and straw storage areas are shown separated from the remaining buildings except for a covered way. This fire break has the recommended minimum width of 15 ft. and to retain this as a fire break the roof of the covered way is of 1 hour's fire resistance. As an added safeguard the end wall of the feed room is also of 1 hour's resistance. The two doors under the covered way, one to the stores and the second to the feed room, are made self closing and are of 1 hour's fire resistance.

In this scheme 28 horses are stabled in one building. To minimise both the risk of and the possible spread of a fire within this building it is divided into four main compartments each of one hour's fire resistance. The walls shown are constructed to 1 hour's fire resistance. The end walls and the roofs over the main walls are made fire resistant for at least 3 ft. on each side of the main division. The doors within the walls forming these sub-divisions are made self closing and of equal resistance.

The main escape routes from the boxes are indicated and alternative exits are provided to each sub-division. The doors in these escape passages are a minimum of 6 ft. wide and open outwards in line with the escape. Those opening into the central passage are set back so that they do not restrict the width of the passage when open. The doors to the boxes are shown opening into the line of main escape route and all close back against the partitions.

Four of the recommended fire points are shown, two to serve the stables and two to serve the hay, feed and straw stores. Stand pipes are situated close to each point outside the buildings. Alternative stand pipes are situated inside the buildings for normal watering use and may be brought into use if required. It is considered that on a scheme of this type and size that the fire points shown are not sufficient protection and two hydrants with reels, as already recommended, are indicated to the north side of the main stable building. These are positioned to serve both the stable building and the hay, feed and straw stores.

When a scheme incorporates a garage for the motor horse box a suitable extinguisher should be fixed adjacent to this garage. The number, type and capacities of such extinguishers should be agreed with the fire brigade.

This layout is close to the main road (not shown) and the whole of the buildings are within the required distance for the requirements of the brigade. It is not therefore essential to allow for suitable roads for the brigade's vehicles within the site in this case.

WATER SUPPLY

The water supply to most stables will be provided by the local water authority and their regulations must be adhered to. In many cases, however, these regulations will not cover the needs of stable buildings in detail and each case will be considered on its merits. It is therefore impossible to give any definite guidance on water supply as the regulations throughout the country vary; and of course the size of the establishment and the type and number of fittings installed will all have a variable effect on the details of the final installation. In all cases the appropriate authorities should be approached as soon as details of size, fittings, fire hydrants, stand pipes, etc., have been agreed with the clients. In the case of fire hydrants and fire points agreement must first be reached with the brigade. This information will enable the authority to give full details of their requirements in respect of the scheme.

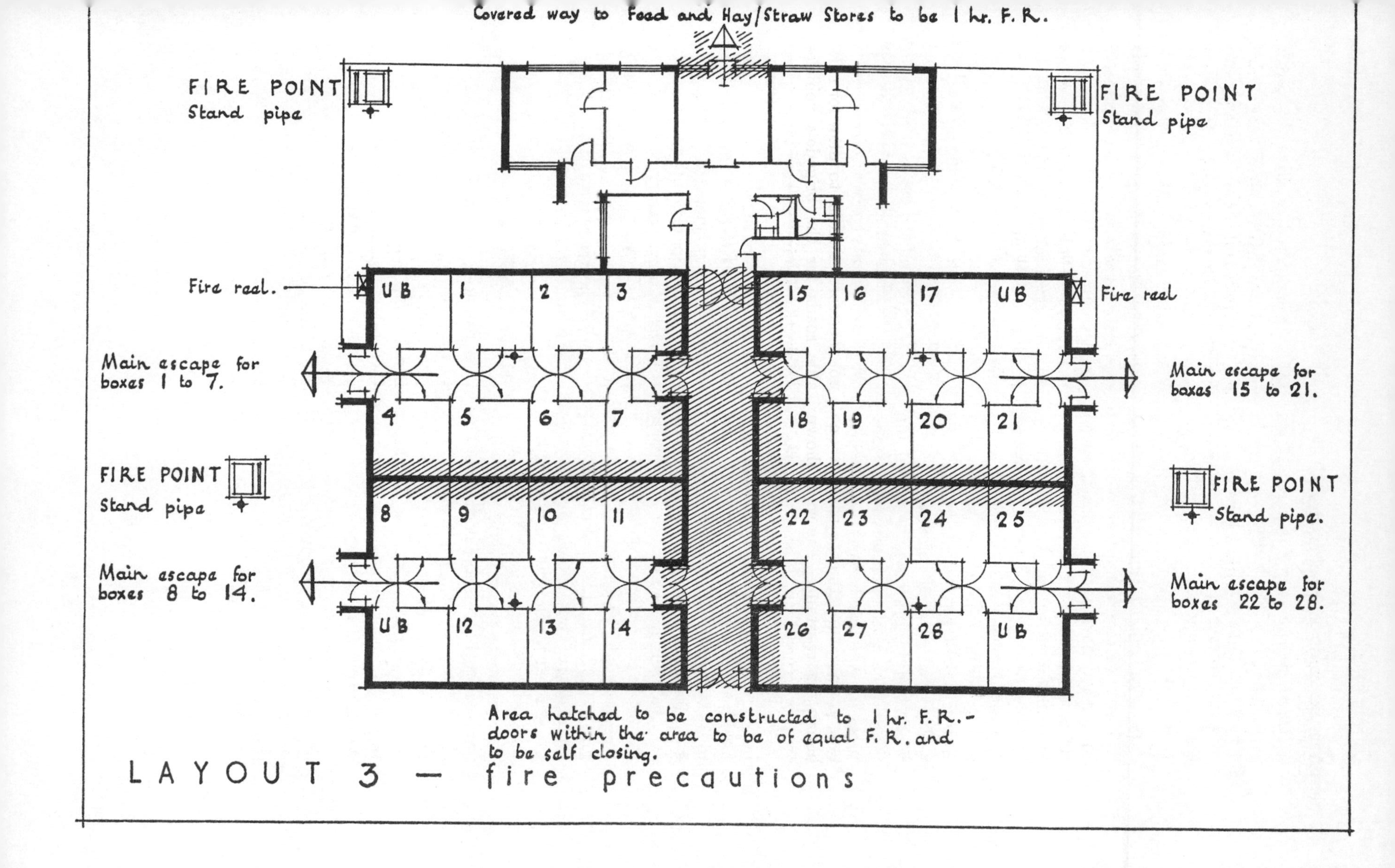

LAYOUT 3 — fire precautions

There are a number of factors relating to water supply which are individual to stable buildings and these must be born in mind during the detailed planning and the preparation of the specification as follows:

(*a*) Storage Cistern and Supply. The storage cistern must be positioned so that any noxious fumes arising either from the boxes or other parts of the buildings cannot taint the water. It has already been pointed out that horses will often refuse to drink fouled water and it will be found in many cases that water tainted by the fumes from the stable will also be objectionable to them. Cisterns must also be lagged and fitted with a dust proof cover. Stand pipes for watering horses should, whenever allowed by the authority, be taken off the "rising" main.

(*b*) Pipes. Pipes where exposed in boxes, or anywhere where there is a likelihood of interference by the horses, must be run at high level, not less than 8 ft. from the floor. Where it is necessary to bring pipes to lower levels they should be chased into the walls and such chases should be covered. The covers must be of adequate strength to withstand damage and should preferably be fitted flush to the wall surface. If it is impossible to form chases for the pipes, they must be covered with heavy galvanised steel covers and each should be shaped to prevent a horse obtaining a purchase on it with its teeth. Any covers to either surface pipes or to chases below 4 ft. must in addition be strong enough to withstand kicking.

Where pipes are required to serve automatic drinking troughs it is recommended that the supplies be carried in ducts beneath the floor and risers fitted to serve the troughs. This method will to a great extent obviate the necessity for bringing any pipes down the walls of the boxes. These risers must, however, be protected as discussed above.

Unprotected cold water pipes must not be run above saddlery or blanket airers. Any condensation on such pipes will drip onto the saddlery or blankets and will stain and spoil them.

(*c*) Stop Cocks. Stop cocks fitted in the boxes must be adequately protected against damage. It is best to position them outside the boxes but if of necessity any are fitted internally they should be fixed in a duct or in a C.I. cock pit in the floor and fitted with a hinged cover.

(*d*) General Considerations. It is recommended that all pipes and water supply apparatus should be effectively protected against frost. Although the boxes are warm when occupied, there are many occasions, i.e., in a hunting establishment, that they are left empty for many hours. The heating in the tack and cleaning rooms may in many stables be intermittent, it is wise therefore, to cater for the conditions prevalent in an unheated building.

The above notes outline the main considerations necessary to incorporate in the specification of the scheme in addition to the appropriate requirements of the water authority.

THE ELECTRICAL INSTALLATION

It is essential that adequate care and forethought should be given to the design and specification of the electrical installation, in all the buildings forming the stable group. A sub-standard installation will be dangerous to both the stable staff and to the horses; at the same time it will constitute a very considerable fire risk. The Riding Establishments Act 1964, shows appreciation of these risks but unfortunately gives no recommendations in respect of the installation. New projects will require the electrical installation to comply with and pass the test of the local electricity board and possibly the insurance company. To pass these tests will not necessarily imply, however, that the installation is satisfactory stablewise. There are many factors individual to such buildings which need to be taken into account and are not covered, so far as is known, by the regulations of any institution or electricity board.

It is not intended to give a full specification in this chapter for the installation in stable buildings, but to explain where normal design details need to be varied, and where certain precautions must be taken to ensure that risks to all parties are reduced to a minimum. To facilitate reference the following notes have been arranged under the normal specification headings.

Installation

It is assumed throughout this chapter that the installation will be carried out in full accordance with the regulations issued by the Institute of Electrical Engineers for the Electrical Equipment of Buildings (at the time of writing the 13th edition 1955, with any subsequent amendments, is applicable). It must also conform to the requirements of the appropriate electricity authority, and the insurance company.

Service

In large establishments a separate intake room will probably be required, but in small establishments the intake cable will be brought into one of the rooms or stores forming a part of the main group. If an office is included in the layout it will serve as a suitable position for the intake cable and the meter, switch and fuse gear. In no circumstances must the cable be brought in, in the area containing the boxes or stalls, or in other areas of possible high humidity, i.e., cleaning room or feed room. Areas containing materials of high fire risk, i.e., hay and straw stores must also be avoided.

Switch and fuse gear should be protected and is best enclosed in a dust proof cupboard unless a separate intake room is provided. Intake rooms and also cupboards, should be of one hour's fire resistance.

In most cases the wiring to lighting and power circuits will be on a single phase. Where a large amount of electrically operated machinery is installed some of the equipment may require three phase working. Details of all electrical equipment and machinery should be obtained from the manufacturers at an early stage in the job, both from the point of view of phasing and so that cable sizes can be agreed with the local board. In some country districts it may be found that the full required supply is not available so it may be necessary to reduce the amount of the machinery installed or allow for the provision of a transformer house. It will be appreciated that the latter provision can materially increase the cost of the scheme.

Wiring and conduits

Wiring generally should be carried out in PVC cables in conduits. In cases where wiring is of necessity run close to heating pipes or other sources of permanent heat it should be in heat resisting cables. Mineral insulated copper sheathed cables may be used in some places but must not be used in areas where the horses are liable to interfere with them.

Conduits must in all cases be of a heavy gauge screw-welded type to comply with B.S.31. In the parts of the buildings subject to humidity, e.g., loose boxes, cleaning room, feed room, lavatories, etc., they should be galvanised inside and out. All junctions must be made with similar gauge boxes.

Conduits, where possible should be

chased into the walls. This is particularly important in the loose boxes and in any parts of the buildings where interference by the horses is likely. If it is not practical to chase them in the boxes they must be run at high level (at least 8 ft.). At lower levels they are liable to damage by a horse either through kicking or biting.

Where conduits are run on the surface they must be secured with spacing saddles at about 36 in. centres, the saddles keep the conduits clear of the walls, thus if any are within reach of a crib biter they will soon be damaged, and may even be pulled from the walls with consequent danger to both staff and horses. In boxes where it is necessary to run surface conduits lower than 8 ft. they must be encased by a heavy galvanised conduit cover suitably formed to prevent a horse obtaining a purchase with its teeth.

In some cases the wiring may need to be run in trunking. This should be of a minimum of 18 s.w.g., Zintex coated sheet steel with enamelled finish. Similar precautions must be taken for the trunking and for its protection as discussed above in respect of conduits.

Earthing

All metal sheathed cables, conduits and switchgear, etc., must be earthed in full accordance with the regulations previously referred to.

Socket outlets

All switched socket outlets should be of 13 amp. rating and be mounted in heavy galvanised boxes of waterproof type to B.S.1363.

Owing to the amount of dirt and dust in stable buildings and the usual method of cleaning being by means of a bucket of water and a mop, no floor outlets must be fitted. An exception may be made in the office, if one is included in the scheme.

Socket outlets for use in connection with the boxes should be fitted outside the box to prevent possible interference by the horse. A convenient height would be a point about 6 in. above the top of the lower leaf of the door. Be careful not to fix a tying up ring outside the box close to the point as this will defeat the object of the advice. Where "Utility" boxes are incorporated in the layout there will be no need of points in connection with the horse boxes. Points required may be fitted in the "Utility" box.

Fused spur units

Where electrical equipment is fitted, e.g., blanket airers, water heaters, etc., fused spur units should be fitted in galvanised iron clad boxes. Each unit should incorporate a 15 amp D pole A.C. switch and a neon pilot light with a red lens. Each cover should be engraved to indicate which piece of equipment it serves.

Fan assisted heating units

All such units will require a fused spur unit as detailed above.

Connections to motors and appliances

No conduit must be connected direct to a piece of apparatus subject to vibration. Conduits in such cases must terminate in a through type conduit box fitted closely adjacent to the apparatus. The cable must then be confined in a flexible metallic tube of the best quality "packed" type to ensure the exclusion of dust, water and oil. This tube must be connected to the terminal box of the apparatus. Such flexibles will require independent earthing by an earth wire running the full length of the tube.

Starters and isolators must be fixed closely adjacent to the apparatus. If several pieces of apparatus are fixed close together each starter must be clearly labelled with an engraved plate to indicate which piece of apparatus it controls.

Switches

It is recommended that all switches, whatever their position, should be of a waterproof pattern. Switches outside boxes should be fixed in a convenient position to the doors and clear of any possible interference by the horse. In most stables the lighting points in the boxes are each controlled by a switch situated outside the box. It is considered better practice to control all lights to boxes from a central point. Where "Utility" boxes are incorporated in the scheme, as in Layout 2, the lights to each group may be controlled from switches placed in the "Utility" box. In schemes not incorporating this method of grouping, the switches should be positioned in the tack room or in the office.

If emergency lighting is installed the switch controlling the system should be labelled "EMERGENCY LIGHTING."

Lamp holders

All lamp holders should be of a Home Office all-insulated pattern.

Light fittings

Waterproof bulkhead fittings are recommended for use throughout the entire stable block. In the boxes they should be positioned at high level well out of reach of the horses. In some cases it may be necessary to place them at lower levels within reach of the horses. The fittings used in such cases should be of heavy-weight pattern with each glass protected by a heavy galvanised steel grille cover. There is little danger of damage either to or from the horses if the heavy pattern fittings of this type available are used, and at lower levels they are more easily cleaned and lamps are more easily replaced.

In hay and straw storage areas ensure that the lighting fittings, whatever type is used, cannot by accident come into contact with the material. A hot bulb if very close to or touching dry hay or straw can very easily cause a fire.

Electric heaters

The heating of the tack, cleaning and feed rooms in most establishments, particularly small ones, will be by means of electric fires or other form of electric heater. All fires should be permanently fixed and be at high level. This recommendation does not apply in cases where off-peak storage heaters are used, but ensure that such heaters are not placed close to saddlery.

Fire alarms

Even in small establishments the installation of an electrically operated fire alarm system is recommended, it is considered essential in a large establishment (see chapter on fire precautions). The system must be kept separate from the main electrical installation and should consist of a trickle charger, battery, fire alarm pushes and bells.

The position for these components must be decided relative to each scheme and are best agreed with the appropriate county fire brigade.

The above notes cover the main essentials of the installation relative to stable buildings. In other respects the normal requirements of the appropriate regulations should be complied with. Care must be taken to ensure that once the system has been properly installed that "amateur" electricians are not allowed to tamper with it in any way. To ensure the continuing safety of the system it is advised that tests and inspections are carried out every two years. Such tests should be carried out by a competent and qualified electrical engineer* and any recommendations given in his report should be promptly dealt with.

* A member of the Electrical Contractors' Association or National Inspection Council for Electrical Contractors.

DRAINAGE

Apart from the necessity to conform to the requirements of the appropriate local authority there are two main problems in respect of the drainage of stable buildings which require to be satisfactorily solved.

(*a*) The contamination of the air within the loose boxes caused by an excess of undrained urine.

(*b*) The frequent blockage of the drainage system.

When considering the contamination of the air it must be appreciated that complete freedom from this cannot be obtained due to the retention of a certain amount of urine in the litter. The amount retained will depend both on the type of litter used and the efficiency of the management. The architectural problem is the efficient disposal of the urine not retained in the litter.

Many stable buildings incorporate herring-bone grooves in the floors of the boxes or stalls. The floors are sloped towards an open channel which in turn discharges into a gully either inside or outside the boxes. This method of drainage is not recommended for the following reasons:

(*a*) The grooves very soon become chipped and in time worn. The floor slope is of necessity slight and these two factors combined tend towards retention of the urine rather than accelerate its dispersal.

(*b*) The channels are often little more than slight hollowings in the floor surfaces and are given a negligible fall. As mentioned above in connection with the floor the tendency is to retain the urine. If the channels are properly formed of glazed fireclay units and protected with grilles the retention of urine is less, but such channels are made of only a slight fall and if of any length they invariably retain a certain amount of urine. It is also difficult to ensure that such channels are kept clear of litter, droppings, etc., which will, of course, encourage the retention.

Earlier in this book during the discussion on types of flooring for the loose boxes, the deficiencies of providing herring-bone grooves was pointed out. The preference then given was for a smooth non-slip surface which would more easily allow the urine to flow to the outlet. It was also recommended that the floor of each box should be sloped to fall directly into a gully within the box. This gully if trapped will retain the urine in the trap, although to a certain extent it will be diluted by water if the boxes are washed down daily, as indeed they should be. If this daily washing down is not carried out the trap will soon retain urine only with no dilution.

To prevent this possibility it is considered better practice to fit an untrapped gully to each box which should be connected by a short branch to a main drain which will collect the branches from a series of boxes. This drain in turn should be connected to either a mud or petrol gully before connection to the main drainage system. The drain should be extended at its upper end as a vent pipe and carried up to high level to discharge into the open air. This method allows full ventilation of the drains to the boxes and minimises the quantity of fumes likely to enter the boxes. It is recommended that a maximum of six boxes only are connected to any one ventilated drain.

The possibility of blockage of the drainage system is a serious problem in all stables. Litter from the boxes will frequently get into the drains local to the boxes and it must be prevented from getting into the main system. The recommended layout described above takes this problem into consideration. Any litter or solid matter getting into the drains from the boxes will collect in the removable tray which is a standard provision with either mud or petrol gullies. This tray may be removed and cleaned at regular intervals. At the same time the drains may be so laid out that the branches from the boxes and the main drain to which they are connected are easily rodded. This method may be further improved by fitting each gully within the boxes with a wire tray. This tray may be removed and emptied each time the boxes are washed down.

Similar types of gullies should be fitted in stable yards or in any other position where straw, etc., is likely to collect. Where the yards are large and several gullies are required for drainage purposes, yard gullies may be used which should be connected to a mud gully before the final connection to the main drainage system is made.

To allow for ease of cleaning all trapped gullies should be fitted with cleaning inlets whether on the soil or surface water system.

All gullies and manhole covers should be of a heavy pattern and should be fitted flush with the surrounding paving. Where gullies are fitted in boxes or in paved areas where the horses are likely to pass they should preferably be of a hinged type which will prevent accidental removal and consequential injury to an animal if it should step in it. It is recommended that all manhole covers in similar positions should be hinged for the same reason.

To save labour costs many stables are experimenting with the use of deep litter. If such a system is intended and the local authority approve, drainage to the boxes may be eliminated. This omission will allow a considerable saving in the initial cost of the job. However, it must be appreciated that experiments of this type may not be permanent and traditional methods, or new ones, which require the use of drainage, may later be resorted to.

Although many clients will not wish to interfere with the architect on the subject of drainage providing the particular problem is satisfactorily solved, some it will be found, will have set ideas relative to certain details. It will be wise therefore, for the architect to discuss the matter fully with the client at an early stage of the job.

It is recommended that the local authority is approached early in the scheme as stable buildings are often dealt with as a special case. There may be special conditions to be complied with particularly in respect of the outfall and it is therefore wise to ascertain the full significance of such conditions before the design has reached an advanced stage.

VENTILATION AND HEATING

This subject has been positioned to follow the chapter on drainage as these two subjects, in relationship to stable buildings are closely linked. No system of ventilation can be satisfactory if the efficiency of the drainage system is inadequate and allows foul air to discharge into the stable. It should also be appreciated that the efficiency of the stable management plays an important part in this factor, as the retention of badly soiled litter, irregular cleaning and skipping out will add to the difficulties of providing proper ventilation. A combination of foul drains and inadequate supervision to ensure cleanliness in the boxes will reduce any normally satisfactory system of ventilation to a level far below that necessary for the health of the animal.

The main requirements of any system of ventilation can be summed up as follows:

(*a*) To discharge from the boxes all foul air and to replace it with fresh air at a controlled rate of change.

(*b*) The rate of change should not be sufficient to cause a draught. In the case of human beings the rate is generally accepted as being a maximum of 2 ft. per second and this will be considered as being equally suitable for horses.

(*c*) The change of air must be balanced throughout the entire cubic contents of the box, otherwise stagnant pockets of air will be left.

(*d*) Incoming air must be clean and fresh so the source of the intake must be carefully considered.

(*e*) The temperature must be controlled. In the case of natural ventilation this is not possible but when mechanical ventilation is installed, the air, during cold weather, may be warmed. It should be appreciated however that the horse possesses a perfect mechanism whereby it adjusts its insulating coat to provide for changes in temperature. Horses are, however, stabled, because by clipping and grooming the insulation adjustment factor is reduced. Blankets and rugs are provided in addition to the stable accommodation to help offset this reduction in insulation adjustment. It is therefore a questionable matter if warming the air passing into the stable by mechanical means has any advantages.

A perfect combination of the above principles is difficult to obtain, whatever the type of system used. Natural ventilation it a very hit and miss method, relying as it does, on conditions which may vary hourly. Mechanical ventilation, although to an extent overcoming the variations of natural conditions can only be fully effective in a sealed building, the air entering and leaving thc building through controlled apertures. Such conditions cannot exist in stable buildings; even if windows are fixed, doors will be opened and left open, thereby upsetting the balance of the system. Allowances for such contingencies will of course be taken into account during the design of the installation but 100% efficiency cannot be guaranteed. In spite of these deficiencies a mechanical system is by far the most efficient of the two methods. Many horsemen are opposed to a mechanical system but in spite of this it is felt that such a system must be included in this book.

The type of plan shown in Layout 3, where a number of open boxes have been designed within a totally enclosed building, is probably the most suitable for a mechanical system. Such a building, designed with the minimum number of external doors, and all windows fixed, will more closely give the required conditions for such a system.

Most stables are designed with a series of individual boxes, each with direct access to the open air. During the day the upper section of the doors will be left open, so that the effect of a mechanical system will be destroyed. Such boxes are, however, usually closed at night, so for a period of about 10 hours in every 24 hours they become virtually sealed units (for the purpose of this paragraph it is assumed that no provision is made for natural ventilation other than by opening the doors), a system of mechanical ventilation could, therefore, be considered for use at night only.

The requirements of fresh air depends on the size of the horse. It is usually agreed that the number of cubic feet of fresh air needed may be obtained from the formula:

$$\frac{E}{P} = D$$

E, equals the number of cubic feet of air produced per hour.

P, equals the permissible CO_2 impurity which is taken as 0.03%.

D, equals the desirable supply of fresh air per hour.

In round figures, heavy draught horses ideally require about 3,500–4,000 cu./ft. per hour, horses weighing 1,000–1,250 lbs. (this weight range covers the average weight hunter and riding horse, which has been mainly considered throughout this book) require 2,500 cu./ft. per hour.

Whatever system is employed, draughts must not be created. The maximum air changes recommended are 6–8 per hour. Taking the lesser figure this entails the complete change of the total volume of air in each box every 10 minutes. Beyond this number of changes there is a risk of chilling, particularly in the case of a sick horse or of a young foal.

The minimum rate of air change to keep the premises fresh should not be less than 3 times per hour, and this should include for the corridor as well as the box.

The minimum air quantity will therefore be as follows:

A box measuring 12 ft. × 12 ft. × 10 ft.	=	1,440 cu. ft.
Add ½ corridor 12 ft. × 4 ft. × 10 ft.	=	480 „ „
		1,920 „ „

Therefore 3 air changes per hour = 1,920 × 3 = 5,750 cu. ft. or air per hour.
Allowing 5% margin;= 6,000 cu. ft. of air per hour per animal.
= 100 cu. ft. of air per minute.

Having outlined the basic principles involved in ventilating systems and detailed the needs of the animal it is now proposed to discuss the different types of ventilating systems considered suitable for the buildings.

Natural ventilation

Any design requires to commence at the basic principles, it is therefore intended to refer to those principles which have been outlined earlier in this chapter in an attempt to devise a suitable form of natural ventilation to boxes. A single box measuring 12 ft. × 12 ft. × 10 ft. high will be taken as the basic unit for this exercise.

(*a*) The change of air within the box must be at a controlled rate. The air inside the box is warmer than the external air due to the heat given off by the horses body, and therefore its tendency is to rise. The rate of rise will depend both on its temperature and on its humidity, neither of which can be controlled in a natural system. Inlets must therefore be at low level and the outlets at high level. The control of the rate of air flow can, in a natural system, be obtained only by the size of the inlets and outlets and the method employed in forming these apertures. The rate of flow will be affected by the number of bends in either inlet or outlet ducts and also by the lining to these ducts, a very roughly lined duct will incur high frictional losses which will materially reduce the velocity of the air.

Thus the basic positions of the inlets and outlets have been ascertained.

(*b*) Freedom from draughts is absolutely essential and in any system entails that the velocity of the air must not exceed 2 ft. per second. This rate can only be controlled by the size and type of the air apertures and their positions in relationship to the animal. Fresh air inlets should not therefore be arranged as a series of direct holes in the external walls thus:

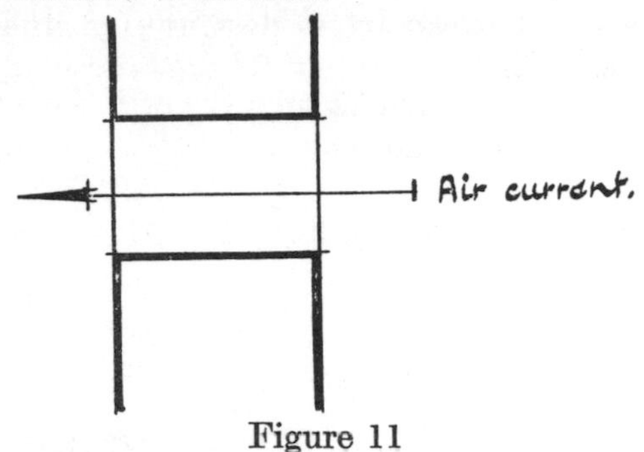

Figure 11

They should be arranged so that the intake is indirect and disperses as quickly as possible as soon as it reaches the inside of the box. There are several methods of arranging this in a simple manner:

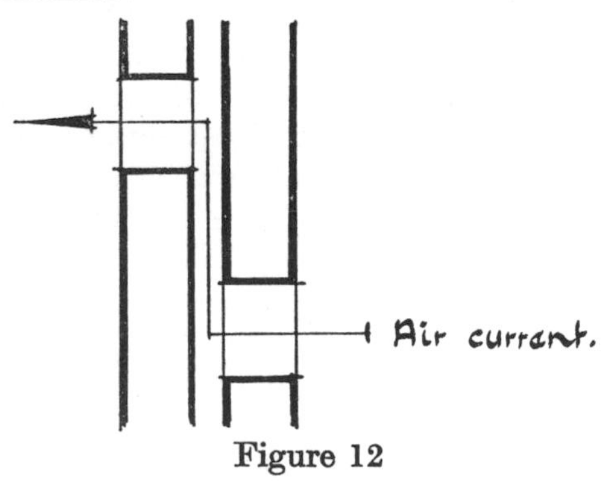

Figure 12

1. By forming a duct within the wall so that the internal and external apertures are not opposite each other.

2. An alternative and better arrangement from the point of view of air dispersal, is to form a duct so that the air enters the box at the side and in an upward direction:

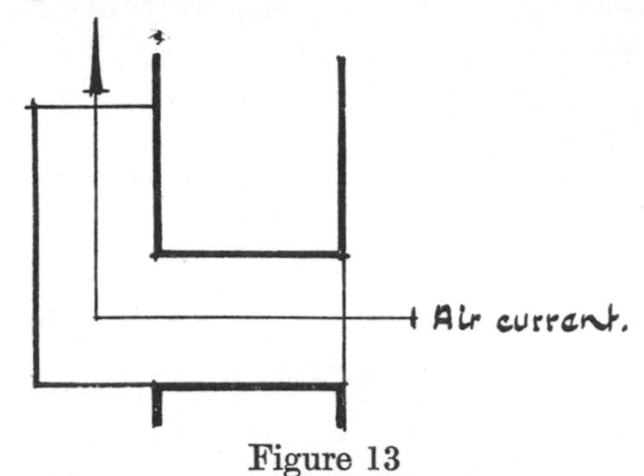

Figure 13

This arrangement eliminates any direct current of air on the horse. It has the disadvantage, however, that a duct of this type projecting into the box may form a hazard to the horse.

3. With the normal practice of cavity wall construction, it is possible to use the whole of the cavity as a duct and so eliminate the objections of the first two methods. To utilise this method it is recommended that the inlets should be fitted at about floor level in the outer skin. The inlets may be arranged in the inner skin in a chequerwise pattern evenly distributed over the surface area of the wall as shown in Figure 14.

The cavity should be sealed at the ends of each wall used for ventilating purposes. It would be preferable if opposite walls of the box were arranged in this manner which would help compensate for wind direction.

(*c*) The balancing of the air changes within the box will depend mainly on the positions of the inlets. If ducts are used as suggested in items 1 and 2 of para. (*b*) it is recommended that at least two are positioned on each side of the box at about one quarter the distance

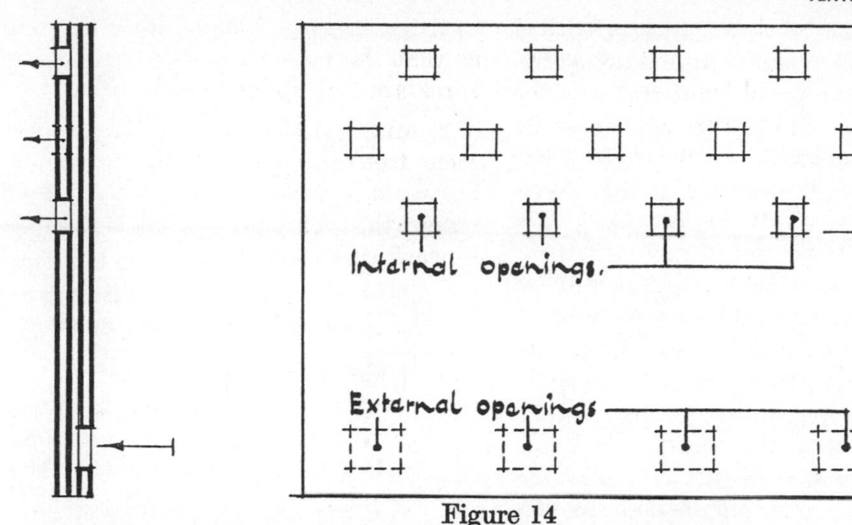

Figure 14

along opposite walls from each corner to give even distribution. The position of the outlets at high level is not so important but in the case of one outlet only being used, it should be fitted centrally in the ceiling. It is considered better practice to have an equal number of outlets to inlets (in this case four), which should be evenly distributed at high level, either through the roof or in the external walls immediately below the highest levels of the ceiling.

In the method discussed in item 3 the outlets should be taken through the ceiling and arranged to give maximum distribution.

(*d*) The requirement of freshness for the incoming air demands care in the positioning of the inlets. Inlets must always be formed directly to the open air but care must be taken that they are not positioned immediately above drainage gullies or other sections of the drainage system likely to give off foul air. They should not be positioned close to manure heaps or in enclosed or restricted areas where the air within such areas is likely to be stagnant.

(*e*) The questionable subject of warmed air does not enter into this section as it is not a practicable proposition to incorporate it in a natural system of ventilation.

The principles discussed above have evolved alternative designs for the ventilation of the box. The outstanding questions are, however:

(*a*) The control of the number of air changes.

(*b*) The control of the velocity of the air.

An old but frequently used formula for calculating the size of outlets and the velocity of the air is S = VAT where,

S equals the cubic content of the air within the box.
V equals the velocity of the air.
A equals the sectional area of the inlet in sq. ft.
T equals the time of air change in seconds.

Taking the basic design unit so far used in this chapter therefore, i.e., a box of 12 ft. × 12 ft. × 10 ft., to find the size of the inlet for an air change of three with a maximum velocity of 2 ft. per second:

$$A = \frac{S}{VT}$$

$$A = \frac{1440}{2 \times 20 \times 60} = \frac{36}{60} = 0.6 \text{ sq. ft.}$$

$$= 86.4 \text{ sq. ins.}$$

If six air changes were allowed the area of the intake would require to be 1.2 sq. ft. = 172.8 sq. ins.

The inlet therefore in the external walls must allow for a free flow of air through them to conform to the above requirements. Allowing for three air changes and an even distribution of inlets, say two to each opposite wall of the box (total four), each must be capable of passing 21.6 sq. ins. of air. Both ends of each duct must be protected to prevent the ingress of vermin and to keep the ducts free from blockage. Louvre type C.1 ventilator grilles or C.1 gratings are recommended. As the size of these outlets will vary depending both on the individual needs of each scheme and on the numbers of inlets provided, a list of suitable types of grilles with their overall sizes and approximate areas of apertures is appended at the end of this chapter. Bearing in mind the basic requirements of the scheme there should be little difficulty in adapting the foregoing information to suit each individual design. Siting will often be the main controlling factor. On an open and exposed site it may be considered better to vere a little to the minimum sizes required for outlets and inlets whereas on a confined site, sheltered by trees or buildings, a slight increase may be made to ensure that the required air changes are as near as possible obtained.

Heavy cast iron gratings with square holes.

9 in. × 3 in. aperture about 5¾ sq. ins.

9 in. × 6 in. aperture about 13½ sq. ins.

9 in. × 9 in. aperture about 20¼ sq. ins.

Heavy cast iron louvres.

9 in. × 3 in. aperture about 4 sq. ins.

9 in. × 6 in. aperture about 15½ sq. ins.

9 in. × 9 in. aperture about 24 sq. ins.

Mechanical ventilation

The main disadvantages of a system of natural ventilation which may be overcome by using a mechanical system can be summed up as follows:

(*a*) However carefully the openings of both inlets and outlets are planned, the results will invariably be upset by high winds.

(*b*) There will be an increase in the quantity of air entering from the windward side which can only be reduced by partly blocking the openings.

(*c*) There may also be a negative pressure on the leeward side to such an extent that the air may be drawn out of the inlet openings, and drawn in through the roof outlet, causing a cold down draught from above.

A plan of the type shown in Layout 3 could with advantage be fitted with a mechanical system of ventilation, and this layout has been used as an illustration of such a system, and is shown facing page 86. The type of system recommended, which it should be noted is of a very simple nature, has been designed with both efficiency and economy of installation and maintenance costs in mind. The illustration clearly shows the arrangement of the system, which consists of an air supply fan, dust filter, air heater with thermostatic control, distributing ducting, supply grilles and fans.

A small plant room has been provided to house the supply fan, heater and dust filter. The dust filter is only necessary to keep the heater fans and ducting from clogging up. A filter such as the glass fibre replacement type would be sufficient. In clean country districts this filter could in fact be dispensed with.

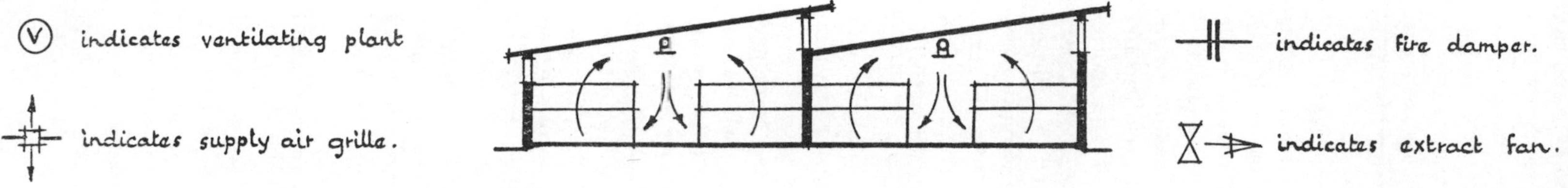

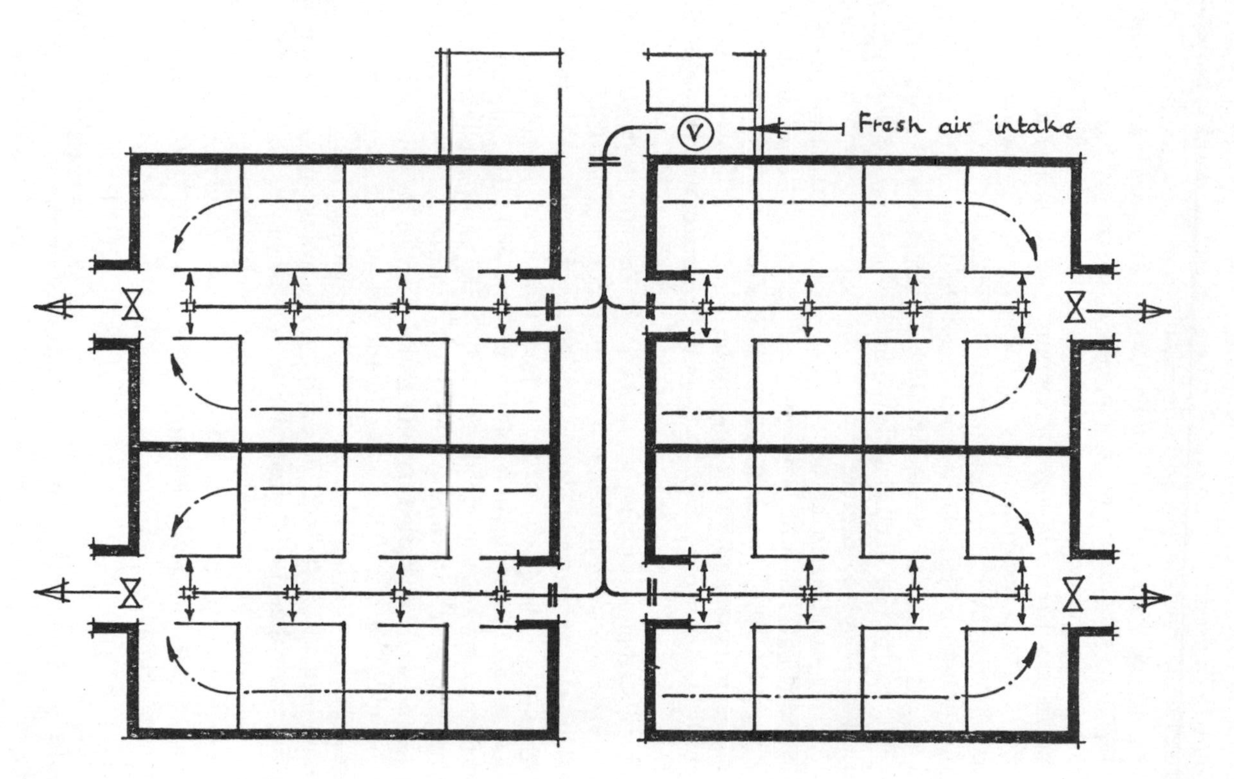

LAYOUT 3. – mechanical ventilation

The air heater could be operated from a low pressure hot water accelerated system using either an oil fired or gas boiler. Alternatively a three-step electric air heater battery could be used. It is anticipated that a temperature of 55 degrees F. in the boxes, when the external temperature is 32 degrees F. would be sufficient. Allowing 100 cu. ft. of air per minute per horse, this works out at approximately 1 kilowatt per horse, so with twenty-eight boxes, an electric heater of approximately 28 kW would be required. The heater would require to be thermostatically controlled and on the scheme illustrated would be best located in the main delivery duct and set at about 58 degrees F.

For this application a cased multi-vane centrifugal fan would be preferred. This would need to handle 2,800 cu. ft. of air per minute, and would need to be 18 in. diameter suction with a 1½ H.P. motor.

The distributing ducting would be of galvanised steel commencing at 24 in. × 12 in. and reducing in size as it goes along. The best positions for the supply grilles would be at high level in the corridors, blowing the air downwards centrally between the boxes. The small air quantities distributed in this manner would not cause draughts in the boxes. The extract fans would be fixed at high level at the ends of the building. In this case four 12 in. propeller fans each handling about 700 cu. ft. per minute would be sufficient. These fans would be fitted with self closing louvres and wind shields.

To overcome fire risks all ducting passing through the fire resisting walls would be fitted with ¼ in. steel plate dampers fitted with fusible links.

PREFABRICATION

Many owners will wish to use both for the stables and the ancillary buildings a form of prefabricated structure. There are many makes and types on the market, the majority of which are constructed in either reinforced concrete or timber. It is not the intention of this chapter to compare any one make with another but only to assist the owner to select a suitable building for his needs from those available.

It is assumed in the first place that the owner has read this book and is therefore fully conversant with the requirements of the units both from the point of view of construction and furnishing. The first consideration to be borne in mind is that most prefabricated structures consist only of the superstructure, so foundations, floors, services, etc., will be required in addition to the structure itself. In many cases these additional requirements will necessitate the employment of an adviser and it will be wise to consult with the adviser at the beginning, as his experience will assist in the selection of the right type of building for the purpose intended.

It is advisable at the outset to inspect as many different makes of buildings as possible. Do not, however, merely study the show models but see the buildings in use. All reputable manufacturers will advise on where their buildings may be seen, so fix appointments, visit the sites and see how each type stands up to normal usage. Discussions with the owners of the buildings should give a fairly clear picture of their advantages and disadvantages. The more examples of each type seen, the more balanced will be the final decision on the one to purchase.

Following the initial research and after the choice has been reduced to two or three makes, it is recommended that before any final decision is made, a visit to each factory is arranged. If this advice is carried out the basic construction of the buildings will be the better appreciated. The soundness of any building is dependent on its basic structure and with many prefabricated buildings the buyer only sees the exterior cladding and the structure itself is taken on trust. The adviser, if one is employed, should also attend the visits to the factories as he will more readily appreciate weaknesses in materials or methods of construction than a layman.

This is also the time to study in greater detail the financial side of the proposed project. Obtain an estimate from each manufacturer and ensure that the full details are given of everything that is required. Read the manufacturers conditions and ensure that their significance is fully understood. In some cases it may be necessary to have certain amendments made to the standard structure. If such alterations are necessary, ensure that the manufacturer can carry them out without any weakening of the building. The cost of any alterations or modifications must be included in the estimate.

Estimates should also be obtained for foundations, floors, drainage, site layout, etc., and also for the erection of the buildings. The combination of these estimates and in addition the cost of fees and expenses will give a complete financial picture of the scheme.

It is appreciated that it is very difficult for either the client or for the architect, if unused to stable buildings, to digest the whole of the recommendations set out in detail in this book and to apply them quickly when studying a prefabricated structure, probably for the first time. To assist in this matter the following list of questions have been prepared, and all will require a satisfactory answer to ensure that the building in question is adequate for the purpose to which it is to be put. To save repetition the questions can be related to any building within the stable group.

Questions

Roof

(*a*) Is the structure adequate in strength?

(*b*) Is the roof covering of suitable material to give adequate protection from the weather and is it satisfactorily secured and laid?

(*c*) What is the life of the roof covering and how soon will maintenance be required upon it?

(*d*) Does the manufacturer provide alternative roof coverings and if so, what are the properties of each and what are the variations in costs between each type?

(*e*) Does the covering or coverings provided by the manufacturer comply with the bye-laws of the local authority?

(*f*) Will the ceiling lining stand up to conditions within the unit in question?

(*g*) Is the ceiling lining satisfactorily secured?

(*h*) Is the ceiling of adequate height?

(*i*) What is the insulation value of the roof and does it equal or approach near to the recommended value?

(*j*) How high is the fire risk, particularly in respect of the roof covering and the ceiling?

(*k*) If a fire started would the overall construction of the roof tend to retard or accelerate the spread of fire?

Walls

(*a*) Is the structure of adequate strength to withstand the aggressive conditions to which it will be subject?

(*b*) Is the internal lining suitable to withstand the conditions imposed in the unit under consideration?

(*c*) Is the lining easily cleaned and maintained?

(*d*) Is the external cladding and finish suitable for the purpose and can it be easily maintained in good condition or does it require frequent treatment to keep a good appearance?
(*e*) Is it adequate in strength?
(*f*) Does the insulation value of the walls equal or nearly so the recommended value?
(*g*) How high is the fire risk of either structure, lining or external cladding?
(*h*) If a fire started would the walls retard or accelerate the spread of the fire?
(*i*) Are the walls suitable and strong enough for fixing the various item of equipment in the unit under consideration?
(*j*) In the case of a sick box is the structure strong enough to support a sling attachment if required?
(*k*) Is the fixing of the walls to the in-situ foundations or floors adequate?

DOORS AND WINDOWS

(*a*) Are the doors adequate in size?
(*b*) Are the doors sufficiently strongly constructed?
(*c*) Do both doors and windows fit properly and are they adequately weathered?
(*d*) Is the joinery of good quality both from the point of view of the quality of the timber and also workmanship?
(*e*) Is the glazing for the windows if provided in accordance with the recommendation?
(*f*) Are the fittings to both doors and windows of good quality and are they sufficient or will additional fittings be required?
(*g*) Are the opening lights to the windows sufficient for ventilation purposes?
(*h*) Are the opening lights placed in a suitable position to provide ventilation without draughts?

FITTINGS

(*a*) If furniture and fittings are provided are they of good quality and satisfactory for the use to which they will be put?
(*b*) Are the fittings for such furniture adequate?
(*c*) If the fittings are provided but not required will the manufacturer credit the fittings not supplied and how much will be the amount of this credit?

GENERALLY

(*a*) What variations can be made to any of the buildings if required and what are the costs involved?
(*b*) Is the ventilation provision suitable to the particular unit in question?
(*c*) What additional fittings, furnishings, etc., are required and what is the cost of such extras?
(*d*) Will the structure pass the bye-laws of the local authorities and if not what additional costs are involved to bring them up to the standard required?
(*e*) Can the structures be satisfactorily joined together to suit the envisaged layout, and site conditions?

Covered Riding School, showing typical construction using standard prefabricated reinforced concrete trusses.

3 COVERED RIDING SCHOOLS, THEIR CONSTRUCTION AND PLANNING

A covered riding school will be required by many establishments and although it may not be considered as an essential part of the stable group it is felt that the book would be incomplete if the detailed requirements of such a building were not included.

This building is intended for use for schooling horses and riders either during inclement weather or when it is considered necessary to keep distractions to a minimum and thus allow horse or rider or both to concentrate with greater ease on the lesson in hand. This requirement is of particular importance when schooling either young horses or novice riders.

Due to the size of the building, which will be discussed later, it should be sited on level ground which should be well drained. If a level site is not available it will be necessary to level over the area of the building and if this is necessary, due care must be taken to ensure that surface water from the upper levels does not drain into the building. Land drains must be laid in such cases to drain the water away from the school. As it will be necessary to excavate to a depth of about 24 in. within the external walls to allow the formation of the floor, it is recommended that any site, whether level or sloping, which has a water table of not less than 36 in. below ground level be drained.

The size of the school will depend upon the needs of the client. A full-size military school measures 60 ft. × 180 ft. This allows for three rides to be instructed at the same time, each working within a square of 60 ft. × 60 ft. Although the individual client's ideas may vary in respect of their own requirements, it is not recommended that this span be reduced as it allows suitable space to longe a horse at the full length of a longeing rein which measures 25 ft. in length. An alternative size which may be in more demand nowadays is 66 ft. × 132 ft. (20 m. × 40 m.) which is the size of a dressage arena. If standard trusses are to be used it may be found necessary to construct the building to the nearest span over 66 ft.

The construction of a building of this size will be best carried out in framed construction using either reinforced concrete, steel or laminated timber. The photograph facing this page shows a recently constructed school formed by using standard reinforced concrete trusses covered with asbestos sheeting. This has proved very satisfactory in use. As insulation factors need not be considered in this building, the architect is left to decide his own method of cladding the framework, which will little doubt be controlled in the main by economic factors on the majority of schools.

The lower parts of the walls on all four sides up to a minimum height of 4 ft. should be of stout construction able to withstand rough usage and kicking. At the same time it must be sloped as shown, at an angle of about 12 to 15 degrees, thus this splay must be formed to prevent either horse or rider from rubbing or knocking themselves against the walls when using the outside track close to the walls. The splay will

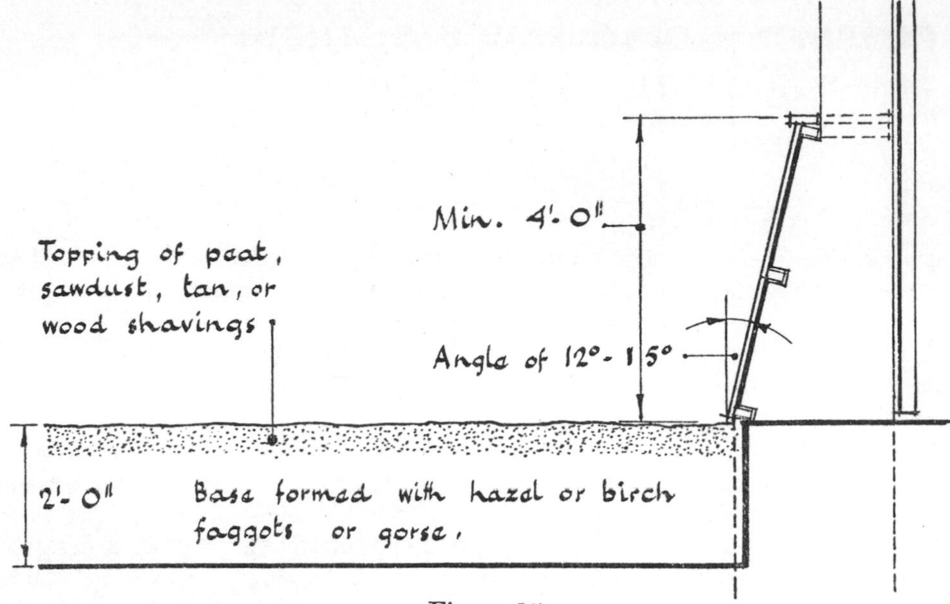

Figure 15

also prevent a horse from rubbing a rider off or damaging his knees. This lower section must therefore be lined through, clear of all post, piers or other projections, and must be continuous and unbroken throughout the length and width of the school. This part of the wall is often constructed of stout timber or of steel framing and faced on the school side with matched boarding. Such construction is sound and suitable for the purpose and will be found to be fairly economical. If non-standard trusses are used the lower sections of the legs may be formed with a splay as shown above in Figure 15. Allowance may then be made for the ties forming the main structure of the splay to be fitted as an integral part of the main framework. Depending on the material used these ties can be formed so that the boarded finish can be secured directly to them. Even if standard trusses are used it will probably prove cheaper and more satisfactory to have purpose-made splayed bases fitted than to form the splayed wall as a separate entity and secure it to the framework after construction. The external cladding to the upper part of the walls should be continued down to ground level to give protection to the back of the splayed lower wall. If economy is not the main consideration and the architect is left a fairly free hand on the use of materials, then there are of course many other satisfactory methods of constructing the building and the architect will then only require to follow the basic design requirements mentioned above.

This building requires good natural lighting with an even distribution of light over the full area of the floor. Some schools have the upper parts of the walls glazed, but this method can form a slight distraction and the insertion of roof lights are generally preferred which will give a more even distribution of light. This matter, however, very much depends on individual preference and the client should be consulted as to his wishes in respect of it.

The floor may be formed by a number of methods. One method is to excavate

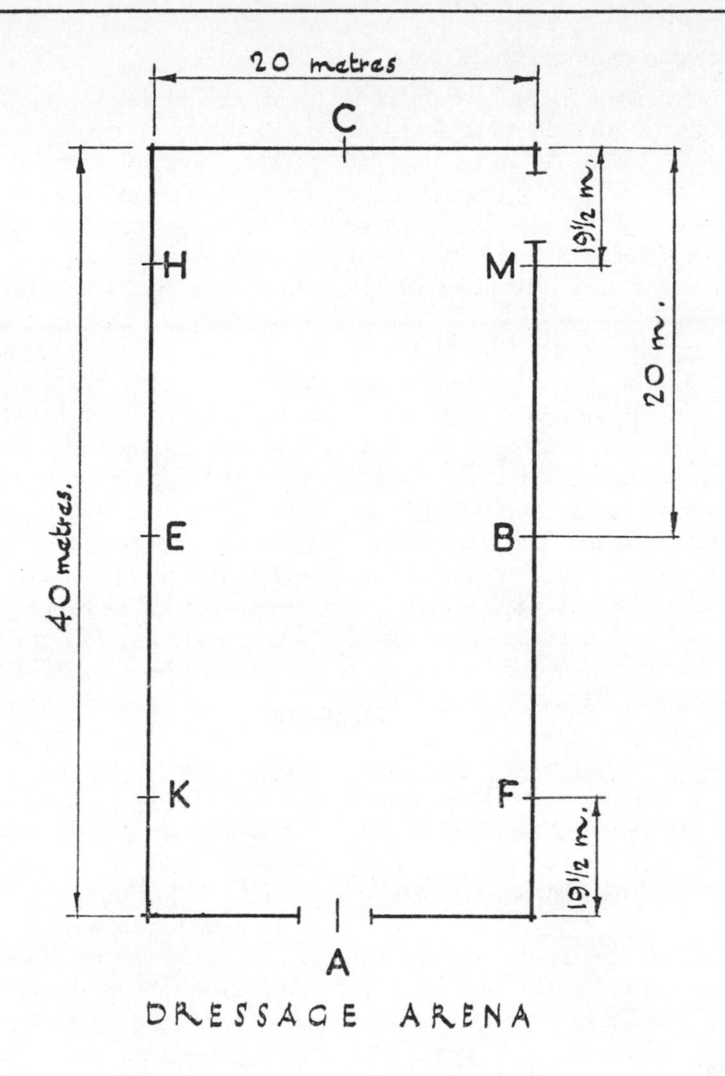

DRESSAGE ARENA

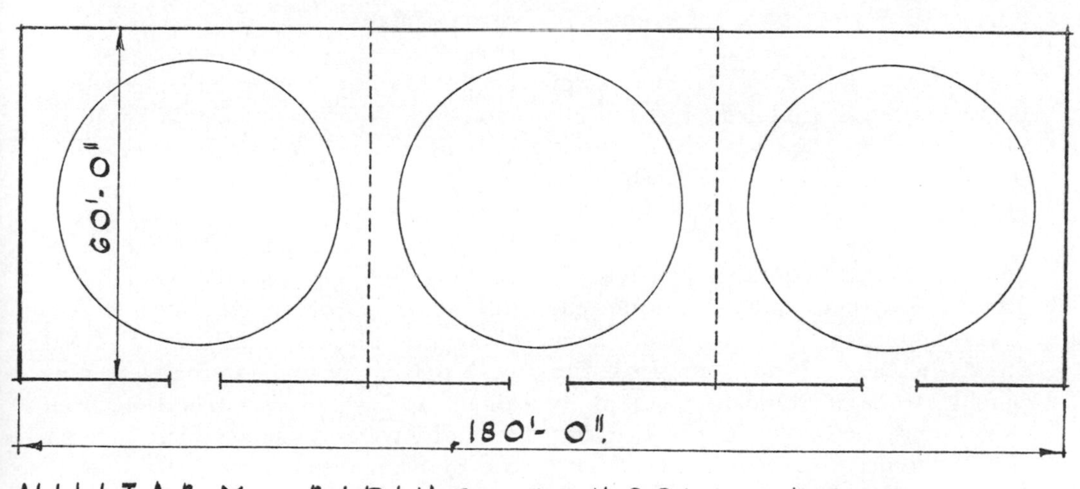

MILITARY RIDING SCHOOL - Full Size

COVERED SCHOOL

the area within the external walls to a depth of 24 in. and fill to a depth of 18 in. with birch or hazel faggots tightly packed. The final 6 in. is then filled with tan. The tan will require to be made up from time to time as it shakes down in the interstices of the faggots.

An alternative to the use of tan is sawdust which has the advantage of not "balling" in the horses feet which is one of the disadvantages of tan. Another disadvantage of tan is that if retained and left in the horses feet it can cause them to heat. The main disadvantage of sawdust and wood shavings is the danger of fire. Peat is probably a more satisfactory finish than either of the foregoing materials and forms a very satisfactory floor. Where economic considerations play an important part in the scheme it would be wise to obtain alternative estimates for the various materials mentioned above before any decision is made as the prices of each type will vary with districts and availability.

The floor of the school shown was constructed with tightly packed gorse instead of the faggots previously described and finished with peat. The original intention was to cover the floor with 6 in. of peat but eventually 4 in. was found to be sufficient.

To give some idea of the amount of peat required the following information kindly given by the owner should prove useful to the designer. The school in question measures 120 ft. × 60 ft.

The peat was bought in compressed bales, each bale containing 8½ cu. ft. Each bale when spread and decompressed equalled about 15 cu. ft. Thus for a thickness of 6 in. over the entire school 240 bales would be required. It was found, however, in practice that slightly under 4 in. was sufficient and 150 bales were used. The gorse base was very tightly packed and the amount of peat lost through shaking down was negligible.

The floor will at times require to be damped. The most economical method of carrying out this operation is to provide a stand pipe adjacent to the building so that a hose fitted with a fine spray may be used or a lawn sprinkler fitted. If economic considerations are not too severe a system of fine shower spray may be fitted beneath the ceiling in such a manner that the sprays give an even distribution of water over the floor. If such a system is fitted ensure that any exposed pipes at low level are adequately protected as detailed for low level pipes in the boxes. It will also be necessary to provide protection against freezing to both pipes and fittings.

The entrance or entrances to the school will depend on site conditions, size and the needs of the client. If a full-size school of 60 ft. × 180 ft. is designed for instructional purposes three doors may be considered desirable, one to each section of the building. A school likely to be used for dressage tests will require a door in the centre of one end wall and a second either at the far end or at the far side of the right-hand wall. This arrangement allows the rider to enter the arena and leave the arena at the correct points.

Entrance doors should be of adequate height to allow a rider to enter whilst mounted. A minimum height of 10 ft. is recommended and a minimum width of 8 ft. Sliding doors are often used and these are satisfactory providing care is taken in choosing a suitable type of gearing. (See recommendations for sliding doors to other buildings in the stable group). Handles should be fitted both externally and internally, one at about 4 ft. high for operation from ground level and one at about 7 ft. for use by a mounted rider.

It is an advantage to provide artificial lighting in this building as such a provision extends its use during the winter months. The installation should be designed to give an overall even distribution of light and this requirement will probably be best obtained by the use of fluorescent fittings. These fittings must be easily accessible as due to the dusty conditions prevalent in this building they will require to be frequently cleaned. Some form of gearing allowing the fittings to be lowered should be considered. The installation should be waterproof throughout and the recommendations included in the chapter on the electrical installation should be followed.

The internal colouring of the school should be light and probably off-white or light grey is the most satisfactory colour scheme. A school likely to be used either as a dressage arena or for dressage training should have half and quarter markers clearly indicated as shown on the illustration facing page 91. The letters must be clear and easily read from any point in the school. The minimum height of lettering should be 12 in. and be in block type of a colour contrasting to the background. Some clients may require a mirror fitted at the centre of either one or both side walls. A framed mirror 10 ft. long × 4 ft. deep would be suitable and it should be secured in a tilting position so that it can be used by a rider on the opposite side of the school. The height should not be less than 6 ft. from the floor level.

Many schools have a gallery formed along one side of the main side walls for viewing, or along both walls. The requirements in respect of such a gallery should be discussed with the client and his full needs ascertained early in the job. This building will in most cases, in the country, be classed by the local planning authority as an agricultural building, but for rating purposes as an industrial building. One small gallery will not vary this classification but if the client requires a large gallery fitted with seating either fixed or movable, the classification will probably be changed to that of a public building. This will entail the normal requirements for such a building to be complied with and can materially add to the expense of the scheme as a whole.

The requirements of the covered school have now been dealt with in adequate detail for the designer to prepare his plans. The outstanding need is that of siting in relationship to the main stable group. The building should be placed in a convenient position to the loose boxes but should be separated from them to prevent horses in the boxes from hearing the work proceeding in the school. Horses at rest very often get upset if the school is placed close to them so that they can hear either the voice of command or the horses at work. The building should also be positioned adjoining and easily accessible from schooling paddocks as it will be used as an extension of these paddocks. A road or paths should connect it to both boxes and paddocks. A school with natural lighting by means of roof lights may be orientated in any direction but one with windows along the sides should preferably be sited due north and south longitudinally and the end walls should be of solid construction throughout.

INDEX